my revision notes

T0173240

CCEA GCSE

HOME ECONOMICS
FOOD AND NUTRITION

Nicola Anderson
Claire Thomson

**HODDER
EDUCATION**
AN HACHETTE UK COMPANY

The Publishers would like to thank the following for permission to reproduce copyright material.

Text acknowledgments
p.66 Information courtesy of the Livestock & Meat Commission for Northern Ireland

Photo credits
p.2 *t* © Stockbyte/Photolibrary Group Ltd/Environmental Issues DV 48; p.2 *c* © Hunta/Fotolia; p.3 © Vangelis Thomaidis/Fotolia; p.4 *tl* © Isa Fernandez Fernandez/Shutterstock; p.4 *tc* © Valentina Razumova/Shutterstock; p.4 *tr* © sss615/Shutterstock; p.4 *bl* © Yuri Samsonov/Shutterstock; p.4 *bc* © baibaz/Shutterstock; p.4 *br* © PR Image Factory/Shutterstock; p.5 *l* © pornpoj/Shutterstock; p.5 *c* © prochasson Frederic/Shutterstock; p.5 *r* © Anastasia_Panait/Shutterstock; p.8 © Colin Underhill/Alamy Stock Photo; p.10 © Crown copyright. Public Health England in association with the Welsh government, the Scottish government and the Food Standards Agency in Northern Ireland; p.11 *t* © Timmary/stock.adobe.com; p.11 *c & b* © Crown copyright. Public Health England in association with the Welsh government, the Scottish government and the Food Standards Agency in Northern Ireland; p.12 & p.13 © Crown copyright. Public Health England in association with the Welsh government, the Scottish government and the Food Standards Agency in Northern Ireland; p.17 © Monkey Business/stock.adobe.com; p.20 © Raphotography/Dreamstime; p.21 *l* © CharlieAJA/Thinkstock; p.21 *c* © kovalchuk/Thinkstock; p.21 *r* © marilyn barbone/Shutterstock; p.22 *c* © Evgenia sh/stock.adobe.com; p.22 *b* © Happy_lark/stock.adobe.com; p.24 *from top* © Zoonar/P. Malyshev/Thinkstock, © PicLeidenschaft/Thinkstock, © Sebalos/Thinkstock, © ibaki/Thinkstock; p.25 *from top* © margouillat photo/Shutterstock, © Wasu Watcharadachaphong/Shutterstock, © bitt24/Shutterstock, © Eivaisla/Shutterstock, © azure1/Shutterstock, © Maks Narodenko/Shutterstock, © marco mayer/Shutterstock, © bitt24/Shutterstock; p.26 *from top* © Aaron Amat/Shutterstock, © Yalcin Sonat/Shutterstock, © Blue Pig/Shutterstock, © Skylines/Shutterstock, © Wasu Watcharadachaphong/Shutterstock; p.28 © CP DC Press/Shutterstock; p.32 © michaeljung/Shutterstock; p.33 © Billion Photos/Shutterstock; p.38 © Rawpixel.com/Shutterstock; p.40 © Shaun Finch - Coyote-Photography.co.uk/Alamy Stock Photo; p.46 © Food Standards Agency; p.49 © Tefi/Shutterstock; p.51 © Crevis/Shutterstock; p.52 *tl* © M. Unal Ozmen/Shutterstock; p.52 *tc* © Maxx-Studio/Shutterstock; p.52 *bc* © Stefano Garau/Shutterstock; p.52 *br* © Wolfilser/Shutterstock; p.59 *t* © Gregory Wrona/Alamy Stock Photo; p.59 *r* © Steven May/Alamy Stock Photo; p.61 © 1000 Words/Shutterstock; p.65 *t* © Fairtrade Foundation; p.65 *b* © The Soil Association; p.66 *t* © Northern Ireland Beef & Lamb Farm Quality Assurance Scheme; p.66 *b* © An Bord Bia, the Irish Food Board; p.73 © Peter Jordan/Alamy Stock Photo; p.77 © stockphotograf/Shutterstock; p.78 © Tomislav Pinter/Shutterstock

Every effort has been made to trace all copyright holders, but if any have been inadvertently overlooked, the Publishers will be pleased to make the necessary arrangements at the first opportunity.

Although every effort has been made to ensure that website addresses are correct at time of going to press, Hodder Education cannot be held responsible for the content of any website mentioned in this book. It is sometimes possible to find a relocated web page by typing in the address of the home page for a website in the URL window of your browser.

Hachette UK's policy is to use papers that are natural, renewable and recyclable products and made from wood grown in sustainable forests. The logging and manufacturing processes are expected to conform to the environmental regulations of the country of origin.

Orders: please contact Hachette UK Distribution, Hely Hutchinson Centre, Milton Road, Didcot, Oxfordshire, OX11 7HH. Telephone: +44 (0)1235 827827. Email education@hachette.co.uk Lines are open from 9 a.m. to 5 p.m., Monday to Friday. You can also order through our website: www.hoddereducation.co.uk

ISBN: 978-1-4718-9933-1

© Nicola Anderson and Claire Thomson 2018

First published in 2018 by
Hodder Education,
An Hachette UK Company
Carmelite House
50 Victoria Embankment
London EC4Y 0DZ
www.hoddereducation.co.uk

Impression number   10   9   8   7

Year   2023

All rights reserved. Apart from any use permitted under UK copyright law, no part of this publication may be reproduced or transmitted in any form or by any means, electronic or mechanical, including photocopying and recording, or held within any information storage and retrieval system, without permission in writing from the publisher or under licence from the Copyright Licensing Agency Limited. Further details of such licences (for reprographic reproduction) may be obtained from the Copyright Licensing Agency Limited, www.cla.co.uk

Cover photo © RTimages/Shutterstock.com

Illustrations by Aptara Inc.

Typeset in India by Aptara Inc.

Printed and bound by CPI Group (UK) Ltd, Croydon, CR0 4YY

A catalogue record for this title is available from the British Library.

Get the most from this book

Everyone has to decide their own revision strategy, but it is essential to review your work, learn key facts and test your understanding. These Revision Notes will help you to do that in a planned way, topic by topic. You can check your progress by ticking off each section as you revise.

Tick to track your progress

Use the revision planner on pages iv–vi to plan your revision, topic by topic. Tick each box when you have:

- revised and understood a topic
- tested yourself
- gone online to complete the quick quizzes.

You can also keep track of your revision by ticking off each topic heading in the book. You may find it helpful to add your own notes as you work through each topic.

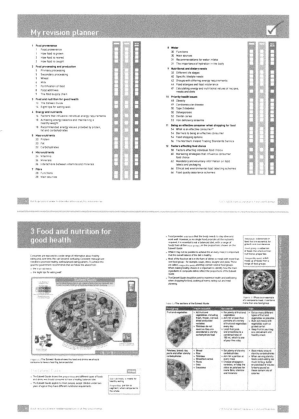

Features to help you succeed

Exam tips

Expert tips are given throughout the book to help you polish your exam technique in order to maximise your chances in the exam.

Typical mistakes

The authors identify the typical mistakes candidates make and explain how you can avoid them.

Now test yourself

These short, knowledge-based questions provide the first step in testing your learning. Answers are given at the back of the book.

Definitions and key words

Clear, concise definitions of essential key terms are provided where they first appear.

Key words from the specification are highlighted in purple throughout the book.

Online

Go online to try out the extra quick quizzes at **www.hoddereducation.co.uk/myrevisionnotes**

My revision planner

Quick quizzes at
www.hoddereducation.co.uk/myrevisionnotes

1 Food provenance

The origin of food, or where food comes from, is referred to as **food provenance**.

> **Food provenance:** the origin of food; where it comes from.

Food provenance

REVISED

- It is not always easy for consumers to know exactly where food comes from, so **food traceability** is important.
- In terms of food provenance, consumers prioritise: poultry, red meat, eggs, fish, shellfish, dairy products, fruit and vegetables.
- Food can be grown, reared or caught.
- It is important that consumers consider **sustainability** when shopping for food.

> **Food traceability:** having access to information about where food comes from and assurances that food is safe to consume.
>
> **Sustainability:** a way of producing and consuming food that protects the economy and environment.

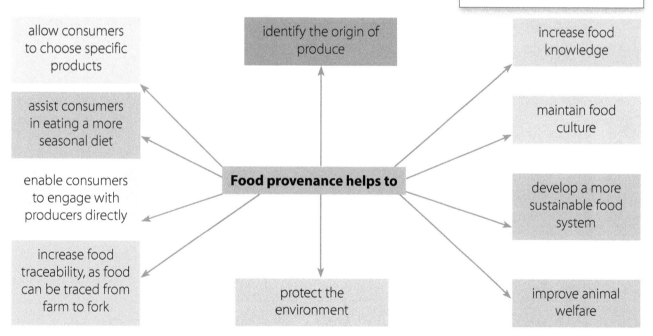

Figure 1.1 **Why food provenance is important**

How food is grown

REVISED

- Farming crops, including cereals, fruit, vegetables, herbs and oilseed, is one way that food is grown in the UK.
- The steps involved in crop production include:
  - preparing soil
  - sowing seeds/seedlings
  - watering
  - fertilising
  - weeding
  - protecting from pests
  - harvesting
  - separation and inspection
  - storage.

How food is reared

REVISED

- Farming animals, including poultry, beef, pork, lamb, goat and game, is one way that food is reared in the UK.
- Farming is a **primary industry** where animals are reared to produce food and other products.
- **Intensive farming** is usually a large-scale operation that prioritises profitability and efficiency.
- **Organic farming** is an alternative type of farming that focuses on producing food in ways that minimise harm to the environment or animals.

Table 1.1 **Intensive and organic farming**

Characteristics of intensive farming	Characteristics of organic farming
<ul><li>High-**yield** crops</li><li>**Pesticides** used to control weeds and pests</li><li>Chemical **fertilisers** used to enrich soil</li><li>Animals kept indoors with limited space</li><li>Mechanised agriculture</li></ul>	<ul><li>Crop rotation linked to seasons</li><li>Hand weeding and natural **pest control**</li><li>Green manure and composting to enrich soil</li><li>Animals are given space to move freely</li><li>Labour-intensive agriculture</li></ul>

Figure 1.2 **Intensive farming often means animals are kept indoors with limited space**

Figure 1.3 **Organic farms allow animals space to move freely**

How food is caught

REVISED

- Fishing for fish and seafood, is one way that food is caught in the UK. There are many different types of fish.

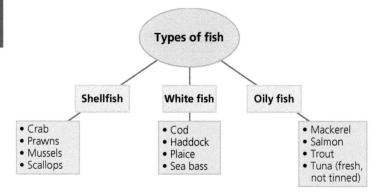

Figure 1.4 **Types of fish**

● A range of methods are used to catch fish.

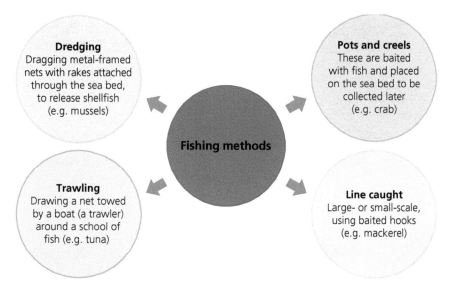

Dredging
Dragging metal-framed nets with rakes attached through the sea bed, to release shellfish (e.g. mussels)

Pots and creels
These are baited with fish and placed on the sea bed to be collected later (e.g. crab)

Fishing methods

Trawling
Drawing a net towed by a boat (a trawler) around a school of fish (e.g. tuna)

Line caught
Large- or small-scale, using baited hooks (e.g. mackerel)

Figure 1.5 **Fishing methods**

Figure 1.6 **Declining stocks of wild fish is one reason why fish farming has become more prevalent**

● Sustainability is an environmental issue associated with fish due to the consequences of overfishing.
● A lack of wild fish is one reason why fish farming (aquaculture) has become more common. Fish can be raised in fish farms using cages, pens, tanks or pods (e.g. salmon, trout, cod, sea bass and rope-grown mussels).

Now test yourself

TESTED ☐

1 Identify **two** examples of food that is reared. [2 marks]
2 Explain **one** reason why consumers may choose to purchase sustainable fish. [2 marks]
3 Outline the main steps involved in crop production. [3 marks]
4 Explain the main characteristics of the following types of farming: [4 marks]
  (a) organic farming
  (b) intensive farming.
5 Discuss why it is important that consumers consider food provenance when shopping for food. [6 marks]

Primary industry: an industry that harvests raw materials from nature, including agriculture and fishing.

Intensive farming: a large-scale operation where profitability and efficiency are prioritised.

Organic farming: a type of farming that focuses on producing food in ways that minimise harm to the environment or animals.

Yield: the produce of a crop.

Pesticide: substance that destroy pests.

Fertiliser: any natural or chemical substance used to make soil more fertile.

Pest control: methods used to prevent or reduce pests, e.g. insects or rodents that would harm crops.

Aquaculture: fish farming.

Exam tip

Make sure key words are spelt correctly and used in context to the question being asked – for example, 'oily' fish or 'organic' farming.

Typical mistake

Many pupils miss the focus of the question and answer with general knowledge rather than subject-specific information. It is important to include relevant examples that show an in-depth understanding of the topic.

2 Food processing and production

Food processing refers to the stages in which raw ingredients are turned into food and made suitable for consumption.

Food is processed to:

- make it safe to eat
- slow down spoilage
- add variety to the diet
- offer convenience to consumers.

Figure 2.1 **Corn can be processed to create many different foods**

A **food production** system has three parts: input, process and output.

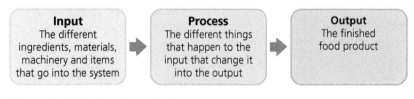

Figure 2.2 **The food production system**

Food processing: the stages used to turn raw ingredients into food and make it suitable for consumption.

Food production: a system that turns raw ingredients into consumable food and drink products.

Figure 2.3 **A food production system for a cake**

Primary processing

● **Primary processing** is when a food has to be processed before it can be eaten – for example, wheat processed into flour.

Secondary processing

● **Secondary processing** is when a food that has undergone primary processing is transformed into a food product – for example, flour processed into pasta.

Wheat

Growing and harvesting wheat

Wheat is grown in a field.	The harvesting process removes the wheat grains from the plant.	The harvested wheat grains are stored until they are needed for use.	The harvested wheat is transported to a mill.

Figure 2.4 **Growing and harvesting wheat**

> **Primary processing:** when a food has to be processed before it can be eaten.
>
> **Secondary processing:** when a food that has undergone primary processing is transformed into a food product.

Processing wheat into bread

Cleaning
The wheat is cleaned and conditioned (to soften the outer bran layer).

Milling
Flour is milled using a range of rollers and sieves.

Flour
The bran, wheatgerm and endosperm are separated and blended into different types of flour: wholemeal, brown and white.

Bread
Secondary processing uses products like flour and converts them into more complex foods, such as bread. To produce bread, the secondary processing steps would include: weighing and measuring, mixing, proving, shaping, baking and slicing.

Figure 2.5 **Processing wheat into bread**

Producing milk

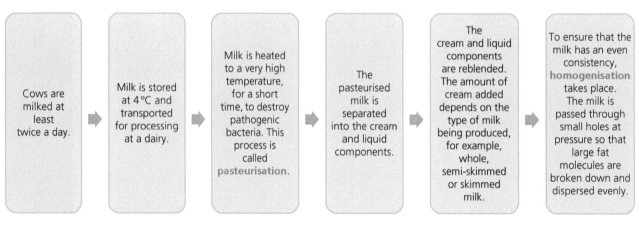

Figure 2.6 **Producing milk**

Processing milk into cheese

A range of different processing methods can be used at this stage, depending on the type of cheese being produced:

- Specific bacteria are used to thicken milk.
- Rennet (an enzyme used to separate milk into curds and whey) is added to milk.
- Curds (solid) are cut and whey (liquid) is released and drained from the curds.

> **Pasteurisation:** a process that reduces harmful bacteria by heating products to very high temperatures and cooling them rapidly.
>
> **Homogenisation:** a process that breaks down large fat molecules and disperses them to improve consistency.

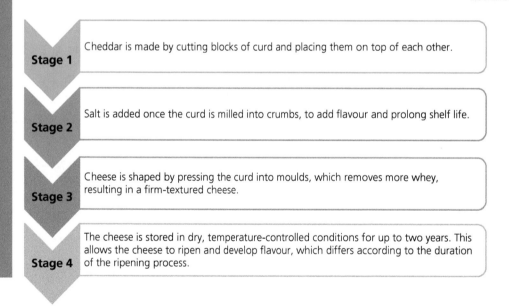

Stage 1 Cheddar is made by cutting blocks of curd and placing them on top of each other.

Stage 2 Salt is added once the curd is milled into crumbs, to add flavour and prolong shelf life.

Stage 3 Cheese is shaped by pressing the curd into moulds, which removes more whey, resulting in a firm-textured cheese.

Stage 4 The cheese is stored in dry, temperature-controlled conditions for up to two years. This allows the cheese to ripen and develop flavour, which differs according to the duration of the ripening process.

Figure 2.7 **Processing milk into Cheddar cheese**

Fortification of food

- **Fortification** of food means that nutrients – usually vitamins and minerals – have been added to food and drink products during processing and production. This is done to ensure that these important nutrients are included in the diet.
- The most common types of fortified products are **staple foods** such as bread, cereals and margarine.
- Fortified foods make an important contribution to the diet of many people in the UK.
- The fortification of food is closely controlled for safety reasons, as the consumption of large amounts of some nutrients could be harmful to health.

> **Fortification**: nutrients added during the processing and production of food and drink products.
>
> **Staple food**: a food regularly consumed as part of the diet.

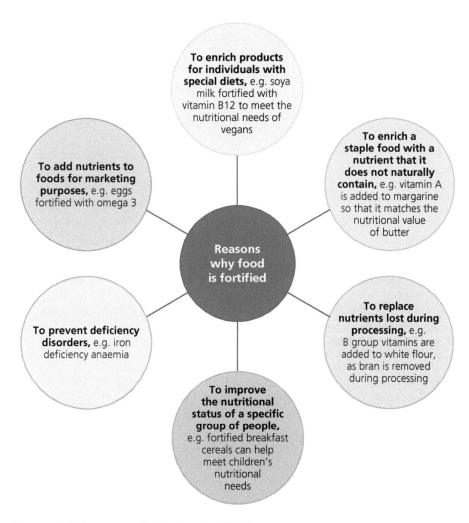

Figure 2.8 **Reasons why food is fortified**

Food additives

- **Food additives** are ingredients that are sometimes added to foods during processing and production.
- Some additives are **artificial**, while others are **natural** or **nature identical**.
- They are added for a variety of reasons, including to enhance flavour or colour, or to increase the product's shelf life.
- Additives must be assessed for safety before they can be used in food and, once approved, will be given an **E number**, which indicates that it has been accepted as safe for use and consumption.
- Additives must be clearly stated in the list of ingredients on a food label.
- Additives can be grouped according to their function.

Food additives: natural or synthetic substances added to food to perform a particular function.

Artificial: synthetic additives made entirely from chemicals.

Natural: additives taken from one food and used in another food.

Nature identical: additives made artificially to be the same as a natural product.

E number: a number allocated to an additive that indicates it has been approved and is safe for use and consumption in the EU.

Figure 2.9 It is a statutory requirement that food additives are clearly stated in the list of ingredients on a food label

Table 2.1 **Common additives and their functions**

Type of additive	Function of additive	Example of foods
Antioxidants	• Extend shelf life • Decrease risk of oils/fats in foods changing colour or going off	• Fruit juice • Salad dressings • Sauces • Bakery products
Colours	• Make food look more attractive • Replace colour lost during processing • Enhance naturally occurring colour of food	• Soft drinks • Fruit yoghurts • Tinned peas • Confectionery
Emulsifiers, stabilisers, gelling agents and thickeners	• Help mix ingredients together and prevent them from separating during storage • Give foods a smooth texture • Give foods a gel-like consistency • Thicken foods	• Jam • Ice cream • Mayonnaise • Sauces
Flavourings and flavour enhancers	• Replace flavour lost during processing • Intensify flavour in food • Add flavour and improve the taste of food	• Sauces • Soup • Chilled meals • Savoury snacks
Preservatives	• Keep food safe to eat for longer • Extend shelf life	• Bread • Biscuits • Dairy products • Cured meats (e.g. bacon)
Sweeteners	• Used with or instead of sugar to make food taste sweet or sweeter	• Low-calorie products • Soft drinks • Yoghurts • Desserts

The food supply chain

- Food goes through a number of processes from the time it leaves the farm until it reaches our fork; this is known as the **food supply chain** (FSC).
- The FSC connects the agricultural, manufacturing and distribution sectors within the food and drink industry.

| The food supply chain usually starts within the **agricultural sector**, where food is grown, reared or caught. | Food goes through the primary and secondary stages of food processing and production within the **manufacturing sector**. | Food is then ready to be supplied to food businesses using a range of different transport methods within the **distribution sector**. | Food and drink products are finally available to consumers from a range of outlets. |

Food supply chain: the processes that food goes through from farm to fork.

Agricultural sector: the part of the food industry where food is grown, reared or caught.

Manufacturing sector: the part of the food industry where food goes through stages of processing and production.

Distribution sector: the part of the food industry where food is transported and supplied to food businesses.

Figure 2.10 **The food supply chain connects the agricultural, manufacturing and distribution sectors**

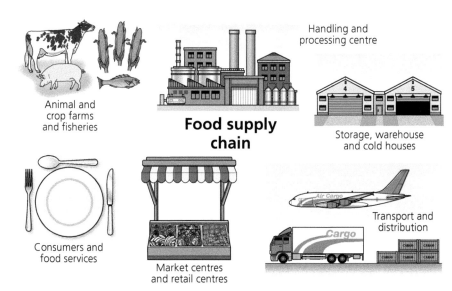

Figure 2.11 **The food supply chain**

Exam tip

Use the command word as a guide to the depth of the topic. For example, 'describe' is a lower order skill than 'explain', so you need to revise a wider range of points for a topic which requires explanation.

Typical mistake

Many pupils mix up the functions and foods linked to the different types of food additives.

2 Food processing and production

Now test yourself

1 Identify **three** of the secondary processing steps involved in making bread. [3 marks]
2 Describe how milk is produced. [4 marks]
3 Outline the food supply chain from 'farm to fork'. [5 marks]
4 Explore the use of additives in food. [6 marks]
5 Explain why foods are fortified. [6 marks]

3 Food and nutrition for good health

Consumers are exposed to a wide range of information about healthy eating and, over time, this can become confusing. Consistent messages are needed to promote healthy, well-balanced eating patterns. To achieve this goal the government recommends that we follow the advice from:

● the Eatwell Guide
● the 'eight tips for eating well'.

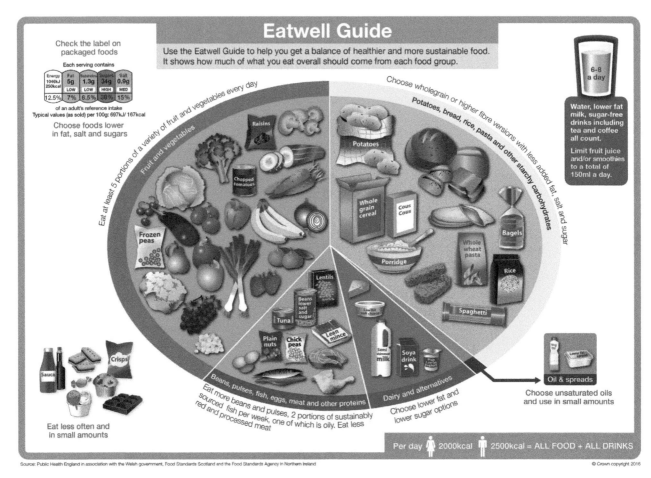

Figure 3.1 **The Eatwell Guide shows the food and drinks we should consume to have a healthy, balanced diet**

The Eatwell Guide

REVISED

● The Eatwell Guide shows the **proportions** and different types of foods and drinks we should consume to have a healthy, balanced diet.
● The Eatwell Guide applies to most people, except children under two years of age as they have different nutritional requirements.

> **Eatwell Guide**: a model for healthy eating.
>
> **Proportion**: portion or segment, when compared to the whole.

- Food provides **nutrients** that the body needs to stay alive and work well. However, as no single food provides all the nutrients required, it is essential to eat a balanced diet, with a range of foods from all five **food groups**, in the proportions shown on the Eatwell Guide.
- While it may not be possible to achieve this at every meal, it is important that the overall balance of the diet is healthy.
- Most of the food we eat is in the form of dishes or meals with more than one food group – for example, soups, stews, lasagne and pizza. These are called **composite meals** and they contain several food groups. When making healthy choices it is important to identify how the main ingredients in composite dishes reflect the proportions of the Eatwell Guide.
- The Eatwell Guide should be used to maximise health and well-being when shopping for food, cooking at home, eating out and meal planning.

> **Nutrients:** substances in food that are essential for growth and maintenance.
>
> **Food group:** a collection of foods that share similar nutritional properties.
>
> **Composite meal:** a dish made up of foods from a range of food groups.

Figure 3.2 **Pizza is an example of a composite meal: it contains more than one food group**

Table 3.1 **The sections of the Eatwell Guide**

Food group	What's included	How much?	Tips
Fruit and vegetables	• All fruit and vegetables, including fresh, frozen, canned, dried and juiced varieties • Potatoes do not count as they are considered a starchy carbohydrate food	• Eat plenty of fruit and vegetables • Aim for at least five portions of a variety of fruit and vegetables every day • Limit fruit juice and smoothies to a combined total of 150 ml, which is one of your five a day	• Eat as many different types of fruit and vegetables as possible • Bulk out meals with vegetables such as grated carrot • Keep fruit in your bag as a convenient and healthy snack
Potatoes, bread, rice, pasta and other starchy carbohydrates	• Bread • Rice • Potatoes • Breakfast cereal • Pasta • Oats • Couscous	• Eat plenty of starchy carbohydrates • Aim for a portion at every meal • Choose wholegrain varieties, or keep the skins on potatoes for more fibre, vitamins and minerals	• Base meals around starchy carbohydrates • When serving starchy foods avoid adding too much fat (e.g. butter on potatoes) or sauces (creamy pasta) as these contain lots of calories

Food group	What's included	How much?	Tips
Dairy and dairy alternatives	• Milk • Cheese • Yoghurt • Fromage frais • Quark • Cream cheese *Creams are not included in this group as they are high in saturated fat; they fit into the 'foods to eat less often and in small amounts' section.*	• Eat some dairy or dairy alternatives • Choose lower fat/sugar options when possible	• Choose semi-skimmed milk • Cheese is high in saturated fat, so buy reduced-fat cheese • Grate cheese instead of slicing it to use less • Use low-fat plain yoghurt as opposed to cream, crème fraiche or mayonnaise
Beans, pulses, fish, eggs, meat and other proteins	• Meat, poultry, game • White fish, oily fish and shellfish (fresh, frozen or canned) • Nuts • Eggs • Beans and other pulses • Vegetarian meat alternatives, e.g. Quorn/tofu	• Eat some beans, pulses, fish, eggs, meat and other proteins • Eat at least two portions (2 × 140 g) of fish each week, one of which is oily • Limit processed meats such as sausages, bacon and cured meats. If you eat more than 90 g per day of red or processed meats, try to reduce the amount to no more than 70 g per day	• When cooking and serving these foods, do not add extra fat or oil • When buying meat choose a leaner cut • Watch out for meat and fish products in pastry, batter or breadcrumbs, as these can be high in fat and/or salt • An 80 g portion of beans or pulses can count as one of your five a day
Oils and spreads	• Unsaturated oils • Soft spreads made from unsaturated oils *Butter is not included in this section as it is high in saturated fat; it is included in the 'foods to eat less often and in small amounts' section.*	• Use these products sparingly as they are high in fat • Cutting down on these types of foods could help to control your weight as they are high in calories	• Choose lower-fat spreads where possible and use sparingly • Check the label and choose oils high in unsaturated fat and low in saturated fat
Foods to eat less often and in small amounts	• Cakes/biscuits • Chocolate/sweets • Puddings/pastries • Ice cream • Jam/honey • Crisps • Sauces • Butter • Cream • Mayonnaise	• These foods are not required as part of a healthy, balanced diet • If included, they should only be consumed infrequently and in small amounts • Most of us need to cut down on the amount of high fat, salt and sugar foods we eat and drink	• Use lower-fat spread instead of butter • Swap cakes and biscuits for a slice of malt loaf or a teacake with low-fat spread • Gradually reduce the amount of sugar added to drinks, e.g. tea and coffee, and food, e.g. cereal

Important diet and health advice is also placed outside the main image of the Eatwell Guide. This includes the advice below.

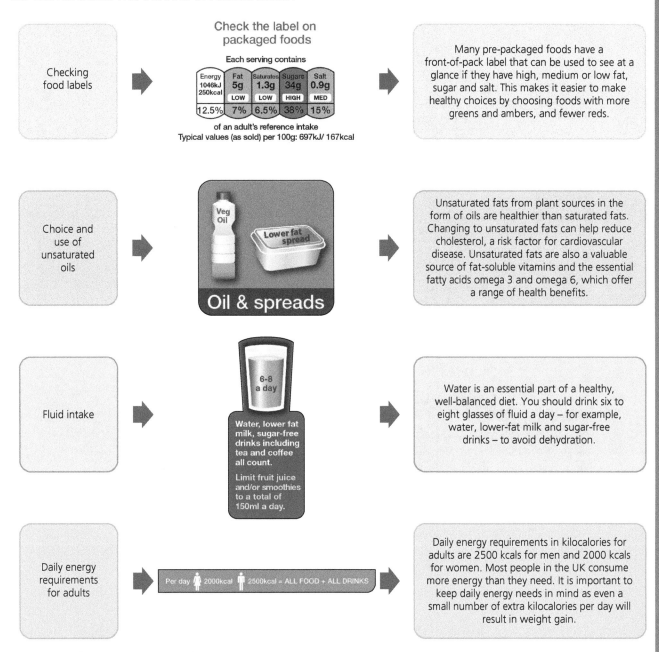

Checking food labels		Many pre-packaged foods have a front-of-pack label that can be used to see at a glance if they have high, medium or low fat, sugar and salt. This makes it easier to make healthy choices by choosing foods with more greens and ambers, and fewer reds.
Choice and use of unsaturated oils		Unsaturated fats from plant sources in the form of oils are healthier than saturated fats. Changing to unsaturated fats can help reduce cholesterol, a risk factor for cardiovascular disease. Unsaturated fats are also a valuable source of fat-soluble vitamins and the essential fatty acids omega 3 and omega 6, which offer a range of health benefits.
Fluid intake		Water is an essential part of a healthy, well-balanced diet. You should drink six to eight glasses of fluid a day – for example, water, lower-fat milk and sugar-free drinks – to avoid dehydration.
Daily energy requirements for adults		Daily energy requirements in kilocalories for adults are 2500 kcals for men and 2000 kcals for women. Most people in the UK consume more energy than they need. It is important to keep daily energy needs in mind as even a small number of extra kilocalories per day will result in weight gain.

Figure 3.3 **Diet and health advice**

Eight tips for eating well

● The 'eight tips for eating well' are designed to help us improve our eating habits and ensure a healthy, balanced diet.

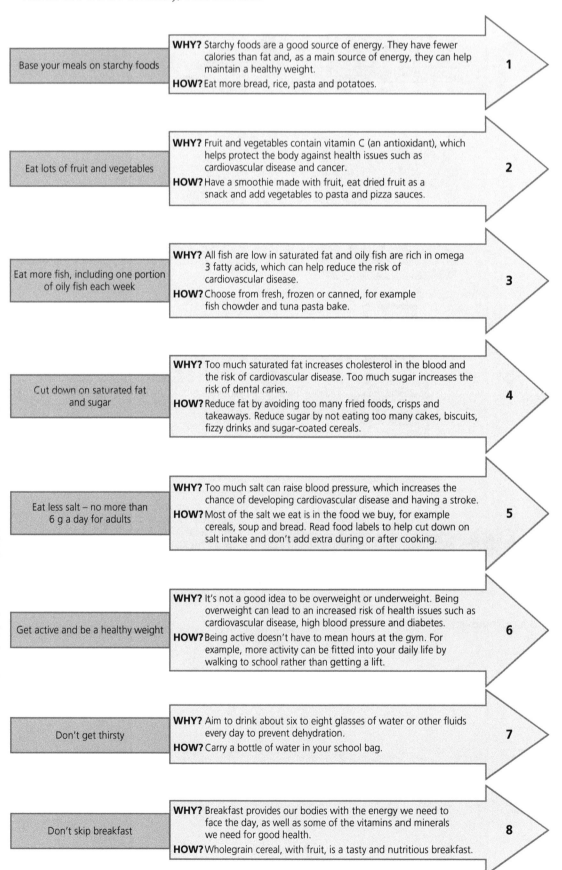

Base your meals on starchy foods

WHY? Starchy foods are a good source of energy. They have fewer calories than fat and, as a main source of energy, they can help maintain a healthy weight.
HOW? Eat more bread, rice, pasta and potatoes.

1

Eat lots of fruit and vegetables

WHY? Fruit and vegetables contain vitamin C (an antioxidant), which helps protect the body against health issues such as cardiovascular disease and cancer.
HOW? Have a smoothie made with fruit, eat dried fruit as a snack and add vegetables to pasta and pizza sauces.

2

Eat more fish, including one portion of oily fish each week

WHY? All fish are low in saturated fat and oily fish are rich in omega 3 fatty acids, which can help reduce the risk of cardiovascular disease.
HOW? Choose from fresh, frozen or canned, for example fish chowder and tuna pasta bake.

3

Cut down on saturated fat and sugar

WHY? Too much saturated fat increases cholesterol in the blood and the risk of cardiovascular disease. Too much sugar increases the risk of dental caries.
HOW? Reduce fat by avoiding too many fried foods, crisps and takeaways. Reduce sugar by not eating too many cakes, biscuits, fizzy drinks and sugar-coated cereals.

4

Eat less salt – no more than 6 g a day for adults

WHY? Too much salt can raise blood pressure, which increases the chance of developing cardiovascular disease and having a stroke.
HOW? Most of the salt we eat is in the food we buy, for example cereals, soup and bread. Read food labels to help cut down on salt intake and don't add extra during or after cooking.

5

Get active and be a healthy weight

WHY? It's not a good idea to be overweight or underweight. Being overweight can lead to an increased risk of health issues such as cardiovascular disease, high blood pressure and diabetes.
HOW? Being active doesn't have to mean hours at the gym. For example, more activity can be fitted into your daily life by walking to school rather than getting a lift.

6

Don't get thirsty

WHY? Aim to drink about six to eight glasses of water or other fluids every day to prevent dehydration.
HOW? Carry a bottle of water in your school bag.

7

Don't skip breakfast

WHY? Breakfast provides our bodies with the energy we need to face the day, as well as some of the vitamins and minerals we need for good health.
HOW? Wholegrain cereal, with fruit, is a tasty and nutritious breakfast.

8

Figure 3.4 **Eight tips for eating well**

Typical mistake

Responses on healthy eating can be vague and lack subject-specific information. Phrases such as 'good for you' should be avoided. Explanations backed up with nutritional evidence are a better way to secure marks, for example:

Fruit and vegetables are a healthy choice as they are rich in vitamin C, an antioxidant nutrient that reduces the risk of diseases such as heart disease and cancer.

Exam tip

You need to be accurate when recalling each of the five food groups and the eight tips for eating well.

You should be able to identify a range of foods in each food group.

You should be able to explain the eight tips for eating well and describe how each one can be achieved.

Now test yourself

TESTED ☐

1 According to the eight tips for eating well, how much salt should adults be consuming every day? [1 mark]
2 List **two** foods found in the oils and spreads section of the Eatwell Guide. [2 marks]
3 Identify the main nutrients supplied by each of the five food groups from the Eatwell Guide. [5 marks]
4 Outline why eating more beans and pulses is recommended for a healthy diet. [2 marks]
5 Why should we be eating less red and processed meat? [3 marks]
6 Explain how the following advice from the eight tips for eating well can help us achieve a balanced diet. [4 marks]
   (a) Don't get thirsty.
   (b) Eat lots of fruit and vegetables.
7 Discuss the Eatwell Guide as a tool for achieving a balanced diet. [9 marks]

4 Energy and nutrients

The body requires **energy** for a range of functions that sustain life.

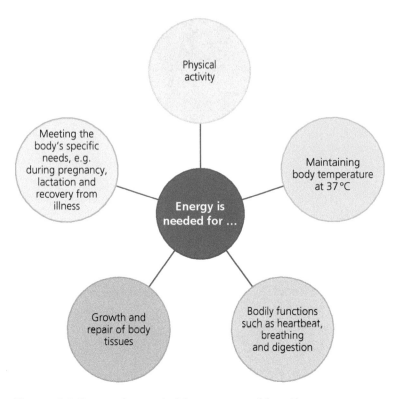

Figure 4.1 **Energy is needed for a range of functions**

Factors that influence individual energy requirements

REVISED

The following factors have an effect on an individual's energy requirements:

- **Basal metabolic rate (BMR)** – this is the amount of energy your body needs to maintain functions such as breathing, heartbeat and body temperature when totally at rest (for example, when sleeping). This accounts for 75 per cent of a person's energy needs.
- **Physical activity level (PAL)** – this is the amount of extra energy needed to carry out physical activities; people who exercise regularly, or who have more physically demanding jobs, need more energy. The impact of PAL on BMR depends on the type and amount of physical activity undertaken.
- **Age** – during childhood and adolescence, energy requirements increase to meet the demands of growth and development. Ageing reduces energy needs, as growth has stopped and levels of physical activity may decline.

Energy: the fuel our bodies need to stay alive and carry out physical activity.

Basal metabolic rate (BMR): the amount of energy the body needs to maintain functions such as breathing and to keep at a constant temperature when totally at rest.

Physical activity level (PAL): the amount of extra energy needed to carry out physical activities, e.g. sitting, standing, walking, running, training or playing sport.

- **Gender** – given that men often have a larger body size, increased muscle mass and a higher BMR than women, adolescent and adult men require more energy to move.
- **Specific need** – women's energy requirements increase slightly to meet the demands of pregnancy and breast feeding (lactation).
- Thermogenic effect **of food** – this refers to an increase in energy expenditure after eating, while the body is digesting food.

> Thermogenic effect: an increase in energy expenditure after eating, while the body is digesting food.

Figure 4.2 **Age and gender both affect our energy requirements**

Achieving energy balance and maintaining a healthy weight

- To maintain a healthy body weight, it is essential to balance energy intake from food with energy expenditure.

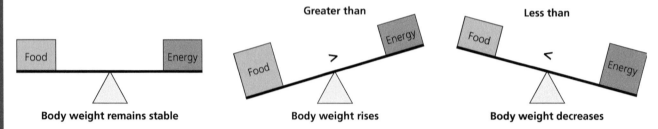

Greater than

Less than

Body weight remains stable

Body weight rises

Body weight decreases

Figure 4.3 **Balancing the amount of energy**

- A number of factors influence energy balance.

Food intake

It is important to consume foods that are nutrient dense (usually high in water) rather than energy dense (usually high in fat and sugar) to meet energy needs; taking in unnecessary calories will lead to weight gain.

Portion size

All food contains calories. Even if you have made sensible choices, you may still be consuming too many calories if you are unsure what a healthy portion looks like. As a result, many people overestimate serving sizes while underestimating calorie content. Over time, this leads to weight gain.

BMR

BMR is affected by a number of factors and it is important to consider these when trying to achieve energy balance. For example, a decrease in BMR in older adults means that energy requirements decrease, which should correspond with reduced energy intake to avoid weight gain.

PAL

PAL is used to calculate the amount of extra energy needed to carry out physical activities (e.g. sitting, standing, walking, running, training or playing sport). PAL will be different depending on the type, amount and intensity of physical activity.

$$PAL = \frac{\text{total energy expenditure (over 24 hours)}}{\text{BMR (over 24 hours)}}$$

Figure 4.4 **Factors that influence energy balance**

Recommended energy values provided by fat, protein and carbohydrates

- Fat, protein and carbohydrates provide the body with energy. The amount of energy in food is measured in **kilocalories (kcal)**.

Kilocalories (kcal): a unit of food energy.

1 gram of **FAT** provides **9** kilocalories

1 gram of **PROTEIN** provides **4** kilocalories

1 gram of **CARBOHYDRATE** provides **3.75** kilocalories

Figure 4.5 **Recommended energy values**

- There are government recommendations for the percentage of a person's energy that should come from these nutrients.

Table 4.1 **Dietary reference values (DRVs) for carbohydrates, fat and protein as a percentage of energy intake**

	% daily food energy
Total carbohydrates of which free sugars	50% Not more than 5%
Total fat of which saturates	Not more than 35% Not more than 11%
Protein	15% (as a secondary source of energy)

Estimated average requirements

- An **estimated average requirement (EAR)** is an estimate of the average amount of energy, or nutrient, required by people in a particular age group.
- EARs are based on the average energy required for people of a healthy weight who are moderately active. These will differ depending on size and gender.

Exam tip

You should be able to explain why people's energy requirements differ and how this can have an impact on food choice.

Typical mistake

Do not mix up the recommended energy values provided by protein, fat and carbohydrates.

Estimated average requirement (EAR): the average amount of energy, or nutrient, required by people in a particular age group.

Now test yourself

TESTED

1. Complete the following statement: 1 gram of carbohydrate provides _____ kilocalories. [1 mark]
2. Identify **two** foods low in energy (kilocalories). [2 marks]
3. Explain **two** ways that older adults can achieve energy balance and maintain a healthy weight. [4 marks]
4. Analyse the data in the table below and answer the questions that follow:

Table 4.2 **Estimated average requirements for adults**

Age (years)	Male (kcals)	Female (kcals)
19–24	2,772	2,175
25–34	2,749	2,175
35–44	2,629	2,103
45–54	2,581	2,103
55–64	2,581	2,079
65–74	2,342	1,912
75+	2,294	1,840

  (a) Identify the age and gender with the highest energy requirement. [2 marks]
  (b) Outline what happens if people continue to eat the same number of calories at 65 as they did at 25. [2 marks]
  (c) Explain how male and female energy requirements vary. [3 marks]
5. Explain a range of factors that influence energy requirements. [6 marks]

5 Macronutrients

Macronutrients are nutrients that are required in large amounts in the diet. Protein, fat and carbohydrates are all macronutrients.

Protein

REVISED

- **Protein** is needed for growth, repair and maintenance of body tissues. It is especially needed to facilitate growth during pregnancy, childhood and adolescence.
- Protein provides a secondary source of energy (1 g provides 4 kcal).

Table 5.1 **Sources of protein**

Animal protein	Plant protein	Novel protein
Meat	Pulses	Tofu
Fish	Beans	Soya
Eggs	Grains	Quorn
Dairy products	Nuts	Tempeh

Biological value of proteins

- Proteins are made up of units called **amino acids**.
- Amino acids that can be produced by the body are called **dispensable amino acids**.
- Amino acids that cannot be made in the body are called **indispensable amino acids**, and must be provided by the diet.
- If a food contains all the indispensable amino acids in the approximate proportions required by humans, it is said to have a **high biological value (HBV)**.
- A food that lacks one or more of the indispensable amino acids is said to have a **low biological value (LBV)**.
- Proteins from animal sources have a higher biological value than proteins from plant sources.
- When there are two LBV protein foods, for example beans on toast, the amino acids in one food will compensate for the limited amino acids of the other, resulting in a meal of high biological value. This is known as **complementation** (the complementary action of proteins).
- Complementation is particularly relevant for vegetarians and vegans.

> **Macronutrients:** nutrients required in large amounts in the diet (fat, carbohydrate and protein).
>
> **Protein:** a macronutrient needed in the body for a range of functions, including growth and repair.
>
> **Amino acids:** units that join together to make proteins.
>
> **Dispensable amino acids:** amino acids which can be produced in the body.
>
> **Indispensable amino acids:** amino acids which cannot be made in the body and which must be provided by the diet.
>
> **High biological value (HBV):** a food that contains all the indispensable amino acids.
>
> **Low biological value (LBV):** a food that lacks one or more of the indispensable amino acids.
>
> **Complementation:** a combination of LBV proteins in one meal to provide all the essential amino acids.

Figure 5.1 **Separately, baked beans and toast each have a low biological value; however, combining them results in a meal of high biological value**

Fat

- **Fat** is used for energy. It provides a concentrated source of energy (1 g provides 9 kcal).
- Fat insulates the body to maintain body temperature.
- Fat protects organs and bones against damage.
- Fat is a source of fat-soluble vitamins A and D, and the essential fatty acids omega 3 and omega 6.
- The body needs some fat to meet its energy requirements. However, no more than 35 per cent of dietary energy should come from total fat, with no more than 11 per cent of this dietary energy coming from saturated fat.
- Fats are composed of fatty acids.
- The body can make all of the fatty acids it needs except for two, known as omega 3 and omega 6. These are called essential fatty acids and must be provided in the diet.
- Omega 3 fatty acids can help prevent blood clotting, which protects the heart and can reduce the risk of cardiovascular disease.
- Omega 6 has a positive impact on blood cholesterol and can reduce the risk of cardiovascular disease.

> **Fat:** a macronutrient needed in the body for a range of functions, including warmth and energy.
>
> **Saturated fat:** fat mainly from animal sources that is typically solid at room temperature; examples include butter and lard.
>
> **Essential fatty acids:** the fatty acids omega 3 and omega 6, which must be supplied in the diet as the body cannot produce them.
>
> **Unsaturated fat:** fat mainly from plant sources that is typically liquid at room temperature; examples include olive oil and sunflower oil.
>
> **Monounsaturated fat:** a type of unsaturated fat with one unsaturated double bond in the molecule.
>
> **Polyunsaturated fat:** a type of unsaturated fat with more than one double bond in the molecule.

Table 5.2 **Sources of fat**

Saturated fat	Unsaturated fat	
<ul><li>Saturated fats are mainly found in foods from animal sources.</li><li>A diet high in saturated fat has been associated with raised blood cholesterol levels, which is a risk factor for cardiovascular disease.</li></ul>	<ul><li>Unsaturated fats are mainly from plant sources</li><li>Many studies have shown that unsaturated fats are better for health</li><li>There are two types of unsaturated fat<ul><li>Monounsaturated fats have one unsaturated double bond in the molecule.</li><li>Polyunsaturated fats have more than one double bond in the molecule.</li></ul></li></ul>	
Food sources: <ul><li>meat</li><li>eggs</li><li>dairy products</li><li>butter</li></ul>	Monounsaturated fat sources: <ul><li>vegetable oils (e.g. olive oil)</li><li>nuts</li><li>olives</li><li>avocados</li></ul>	Polyunsaturated fat sources: <ul><li>oily fish</li><li>meat</li><li>margarine</li><li>vegetable oils</li><li>seeds</li></ul>

Figure 5.2 **Sources of saturated fat**

Figure 5.3 **Sources of monounsaturated fat**

Figure 5.4 **Sources of polyunsaturated fat include meat, oily fish and seeds**

Carbohydrates

- **Carbohydrates** provide fibre, which aids digestion and prevents constipation.
- Carbohydrates provide the body with energy (1 g provides 3.75 kcals).
- Carbohydrates have a protein-sparing effect, so that protein is used for growth and repair rather than energy.
- The body needs a constant supply of carbohydrates to meet its energy requirements. About 50 per cent of dietary energy should come from carbohydrates, with no more than 5 per cent of this dietary energy coming from **free sugars**.

> **Carbohydrates**: a macronutrient needed in the body for a range of functions, including energy.
>
> **Free sugars**: all sugars added to food by the manufacturer, cook or consumer, and sugars naturally present in honey, syrups and unsweetened fruit juices
>
> **Sugars**: simple carbohydrates that are absorbed quickly by the body and raise blood sugar levels rapidly.
>
> **Starches**: more complex carbohydrates that take longer to digest and absorb, keeping blood sugar levels constant.

Table 5.3 **Sources of carbohydrates**

Sugary carbohydrates	Starchy carbohydrates
Sugars are absorbed quickly by the body, and raise blood sugar levels rapidly.	**Starches** take longer to digest and absorb. They help keep blood sugar levels constant. Starchy carbohydrates are also a valuable source of fibre.
Food sources: • table sugar • confectionery • cakes and biscuits • honey • fruit juice	Food sources: • grains • cereals • pasta • rice • some fruit and vegetables

Figure 5.5 **These foods are sources of starchy carbohydrates**

Figure 5.6 **These foods are sources of sugary carbohydrates**

Exam tip

Make sure you know relevant food sources for each macronutrient.

Typical mistake

Questions on macronutrients can often be linked to a specific individual. Make sure you answer in context (see questions 2 and 6 below).

Now test yourself

TESTED

1 Identify the percentage of dietary energy that should come from total fat. [1 mark]
2 Identify **two** good sources of protein that would be suitable for a vegan. [2 marks]
3 Identify **two** foods high in free sugars. [2 marks]
4 Explain what is meant by protein complementation. [3 marks]
5 Explain why it is important to include unsaturated fat in the diet. [4 marks]
6 Explain the importance of starchy carbohydrates in the diet of an adolescent. [4 marks]

6 Micronutrients

- **Micronutrients** are needed by the body in small amounts.
- **Vitamins** and **minerals** are micronutrients.
- Each micronutrient has specific functions and sources.
- Deficiencies of micronutrients can have a serious impact on health.
- Interactions between iron and vitamin C, and between calcium and vitamin D, must be understood in order to maximise health.

> **Micronutrients:** nutrients needed by the body in small amounts.
>
> **Vitamins:** micronutrients essential for good health.
>
> **Minerals:** micronutrients essential for the protection of the body, e.g. bones and teeth.

Vitamins

REVISED

- Vitamins are micronutrients that are needed in very small amounts. Usually only a few milligrams (mg) or micrograms (mcg) are needed per day.
- Vitamins B1, B12, folate and vitamin C are **water-soluble vitamins**. They are found in food and plant sources and are not stored by the body.
- Vitamins A and D are **fat-soluble vitamins**. They are found in foods from animal sources and can be stored by the body.

> **Water-soluble vitamins:** vitamins B1, B12, folate and vitamin C, which are found in fruit, vegetables and grains. They are not stored by the body.
>
> **Fat-soluble vitamins:** vitamins A and D, which are found in fatty foods from animal sources and can be stored by the body.
>
> **Antioxidant:** a substance that reduces the destructive effects of oxidation in the body, protecting cells from damage by free radicals.
>
> **Rickets:** a softening of bones in children, potentially leading to fractures and deformity. The main cause is vitamin D deficiency.
>
> **Osteomalacia:** pain and bone weakness in adults caused by a deficiency of vitamin D.
>
> **Iron deficiency anaemia:** a condition where a lack of iron in the body leads to a reduction in red blood cells.

Table 6.1 Functions, sources and deficiency of vitamins

Nutrient	Functions	Sources	Deficiency
Vitamin A	<ul><li>Healthy eyesight</li><li>Healthy skin</li><li>Healthy immune system</li><li>Normal growth</li><li>Antioxidant that protects cells from damage</li></ul>	<ul><li>Animal (retinol): cheese, eggs, oily fish, whole milk, butter, fortified margarine, liver</li><li>Plant (carotene): carrots, green leafy vegetables (e.g. cabbage), orange-coloured fruits (e.g. nectarines)</li></ul>	

Nutrient	Functions	Sources	Deficiency
Vitamin D	• Regulates the amount of calcium in the body	• Oily fish, eggs, butter, meat, fortified foods (e.g. breakfast cereals) • Most vitamin D is obtained from sunlight on the skin	• In children, leads to skeletal deformity called rickets • In adults, leads to pain and bone weakness called osteomalacia
Vitamin B1	• Assists with release of energy from food, in particular carbohydrates • Promotes normal functioning of the nervous system, muscles and heart	• Fortified cereals • Meat • Wholegrains • Nuts	
Vitamin B12	• Formation of red blood cells • Energy production • Involved in the functioning of the nervous system	• Fortified cereals • Dairy products • Eggs • Meat • Fish	
Folate	• In pregnancy, it reduces the risk of neural tube defects in the baby (e.g. spina bifida) • Involved in the formation of healthy red blood cells • Required for cell division	• Fortified cereals • Green leafy vegetables (e.g. cabbage) • Brown rice • Some fruit (e.g. oranges)	• Can increase risk of anaemia • In pregnancy, may cause neural tube defects in the baby (e.g. spina bifida)
Vitamin C	• Helps the body absorb iron from food, reducing the risk of iron deficiency anaemia • Normal structure and function of blood vessels • Promotes development of connective tissue • Involved in wound healing • Antioxidant, protecting cells from damage	• Some vegetables (e.g. peppers) • Citrus fruit (e.g. oranges) • Berry fruits (e.g. blackcurrants) • Kiwi fruit	• Fatigue • Weakness • Aching joints and muscles • Bleeding gums • Can prevent wounds from healing well

Minerals

- Minerals are micronutrients that the body needs in small amounts. They have many different functions, but in general they are protective and help to keep us healthy.

Table 6.2 **Functions, sources and deficiency of minerals**

Nutrient	Functions	Sources	Deficiency
Sodium	• Helps to keep body fluids in balance • Maintains nerve function	• Table salt • Salty snacks (e.g. crisps) • Processed foods (e.g. processed meat products) • Breakfast cereals • Cheese	
Calcium	• Important for the formation and maintenance of bones and teeth • Necessary for nerve and muscle function • Involved in blood clotting	• Milk, cheese and other dairy foods • Green leafy vegetables (e.g. cabbage) • Fortified soya products • White bread • Fish where bones are eaten (e.g. sardines)	• Can reduce peak bone mass, which is a contributory factor in the development of osteoporosis in later life
Iron	• Needed to form haemoglobin in red blood cells, which transport oxygen around the body • Required for normal energy metabolism • Needed for normal functioning of the immune system	• Haem iron is easily absorbed by the body and comes mainly from animal sources (e.g. red meat, liver, eggs, chicken, fish) • Non-haem iron is not as easily absorbed by the body and comes mainly from plant sources (e.g. pulses, nuts, dried fruit, fortified breakfast cereals, green leafy vegetables, wholegrains)	• Leads to low iron stores in the body and eventually to iron deficiency anaemia

Interactions between vitamins and minerals

Calcium and vitamin D

✔ Calcium is most easily **absorbed** from milk and dairy products.

✔ The presence of vitamin D promotes the absorption of calcium in the body.

✘ Calcium is less easily absorbed from plant foods.

✘ Calcium absorption may be reduced by:

- phytates in plant foods (e.g. cereals and pulses)
- oxalates in vegetables and fruit (e.g. spinach and rhubarb).

Iron and vitamin C

✔ Iron from animal sources, known as haem iron, is absorbed more effectively than iron from plant sources.

✔ Eating food containing vitamin C at the same time as food containing iron from non-haem sources can help the body to absorb the iron. Examples include having fruit juice or fruit with fortified breakfast cereal, vegetables with beans or nuts with rice.

✘ Absorption of non-haem iron from plant sources is reduced by:
- phytates in cereals or pulses
- tannins in tea.

> **Exam tip**
>
> Make sure you know relevant food sources for each micronutrient.

> **Typical mistake**
>
> Avoid repetition and maximise marks for questions on micronutrients. For example:
>
> Explain **two** functions of iron. [4 marks]
>
> This question requires two distinct functions with relevant detail to support your answer, for example one sentence for every mark.

> **Osteoporosis:** a disease characterised by low bone density and deterioration of bone tissue, which results in fragile bones and increased risk of break or fracture.
>
> **Haemoglobin:** the red oxygen-carrying pigment in red blood cells.
>
> **Haem iron:** iron from animal sources.
>
> **Non-haem iron:** iron from plant sources.
>
> **Absorbed:** to be taken up, e.g. the amount of a nutrient taken in by the body.

Now test yourself

TESTED ☐

1 Identify **three** foods in the table that are good sources of vitamin C. [3 marks]

Oranges		Cheese	
Kiwi fruit		Red peppers	
Eggs		White bread	

2 Identify **two** foods with a high sodium content. [2 marks]

3 Explain the effect of a deficiency of vitamin D in the diet. [2 marks]

4 Discuss **two** factors that affect the absorption of iron in the body. [4 marks]

5 Discuss **two** ways the dish below could be changed to increase the calcium content. [4 marks]

Recipe	Ingredients
Chicken with rice and peas	1 chicken breast 1 tablespoon sunflower oil 50 g rice 50 g peas

7 Fibre

Fibre is the non-digestible part of plants. A regular intake of a wide range of fibre-rich foods is essential for good health.

Functions

REVISED

- There are two main types of fibre: **soluble** and **insoluble**.

Table 7.1 **Functions of fibre**

Insoluble fibre	Soluble fibre
• Assists digestion by enabling the body to get rid of waste more effectively • Helps prevent constipation by adding bulk to **faeces**, making them easier to pass • Is filling and can reduce the desire to snack between meals, helping to maintain a healthy weight	• Can help lower blood cholesterol levels, therefore reducing the risk of cardiovascular disease • Helps keep stools soft and easier to pass as soluble fibre dissolves in water and forms a gel in the **digestive system** • Has a positive impact on the control of blood sugar levels

Main sources

REVISED

- It is important to eat a variety of foods that are rich in both types of fibre, to maximise benefits to health.

Table 7.2 **Sources of fibre**

Insoluble fibre	Soluble fibre
• Wholegrain cereals • Wholemeal bread • Wholewheat pasta • Brown rice • Nuts and seeds • Fruit and vegetables	• Grains (e.g. oats, barley, rye) • Pulses (e.g. peas, beans, lentils) • Fruit (e.g. apples, bananas, pears) • Vegetables (e.g. carrots, parsnips, potatoes)

Fibre: the non-digestible part of plants; a type of carbohydrate.

Soluble fibre: a type of fibre that helps control blood sugar and blood cholesterol levels.

Insoluble fibre: a type of fibre that assists digestion and helps prevent constipation.

Faeces: waste matter remaining after food has been digested; it is passed out of the body through the bowel.

Digestive system: the body system responsible for getting food into and out of the body, and for making use of it to keep the body healthy.

Figure 7.1 **Fibre-rich foods**

Exam tip

Try using images of food sources or colour coding to make sure you can differentiate between the two types of fibre.

Typical mistake

Candidates often generalise when answering questions on the functions of fibre and mix up food sources of each type of fibre.

Now test yourself

TESTED ☐

1 Identify **one** health problem that may develop from a low intake of fibre. [1 mark]
2 Write down **two** sources of soluble fibre. [2 marks]
3 State **two** ways that the fibre content of the dish below could be increased. [2 marks]

Recipe	Ingredients
Chilli con carne	1 tablespoon of oil 1 onion 2 garlic cloves 1 tablespoon of chilli powder 1 stock cube 2 tablespoons tomato purée 500 g lean minced beef 400 g tinned tomatoes 250 g white rice

4 Outline **three** ways that an adult could increase the fibre content of their diet. [3 marks]
5 Explain the function of insoluble fibre. [3 marks]
6 Explain how a diet rich in soluble fibre can benefit health. [4 marks]

8 Water

Human beings are unable to survive without water, as it makes up two-thirds of our body mass.

Functions

REVISED

Table 8.1 **Functions of water**

Area of the body	Function
Blood	Water transports nutrients and oxygen through the body in blood
Blood pressure	Water helps to maintain blood pressure
Bowel health	Water helps to prevent constipation
Chemical reactions	Water assists reactions in the body (e.g. digestion)
Excretion	Water helps the kidneys to filter waste, which is excreted as urine
Joints	Water provides fluid to keep joints mobile
Saliva	Water is a key component of saliva, which helps swallowing
Spinal fluid	Water is a component of spinal fluid, which cushions the nervous system
Temperature control	Water helps to regulate body temperature via perspiration
Tears	Water helps to form tears to lubricate the eyes

Main sources

REVISED

- Our daily fluid requirements are met by the food, water and other drinks that we consume.

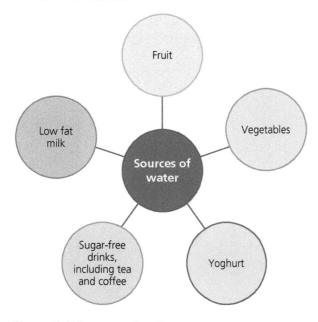

Figure 8.1 **Sources of water**

Recommendations for water intake

- The Eatwell Guide states that we should consume between six and eight glasses of water a day. If each glass is approximately 200 ml, daily fluid intake should be between 1200 ml and 1600 ml.
- One way to achieve this is to always include a drink with meals.

6-8 a day

Water, lower fat milk, sugar-free drinks including tea and coffee all count.

Limit fruit juice and/or smoothies to a total of 150ml a day.

Figure 8.2 **Drink six to eight glasses a day of fluid to avoid dehydration**

The importance of hydration in the body

- Fluid intake is important to achieve good health.
- It is important to develop a habit of regular fluid intake without waiting for thirst to intervene.
- Not drinking enough water can lead to **dehydration**. Symptoms of dehydration include: thirst; headache; dizziness; dry mouth; lips and eyes; overheating of the body; confusion and changes in blood pressure.
- It is essential to drink more water during hot weather to stay **hydrated**.
- Children do not always recognise the early stages of thirst and are less heat tolerant than adults, so they are more susceptible to dehydration in warm climates.
- People should increase water intake before, during and after physical activity to replace fluid lost by sweating.
- As people age, their thirst sensation declines. Older adults should sip water throughout the day and have regular cups of tea/coffee to prevent dehydration.
- Breakfast is also a valuable opportunity to rehydrate – for example, water, fruit juice, tea/coffee and milk on cereal.

> **Dehydration**: a condition caused by the excessive loss of water from the body.
>
> **Hydration**: taking in fluid to restore or maintain fluid balance in the body.

> **Exam tip**
>
> Include a range of relevant key terms when explaining functions of water, e.g. 'excretion'.

> **Typical mistake**
>
> Avoid repetition when responding to questions about hydration: 'preventing dehydration' and 'reduces the risk of becoming dehydrated' are the same point and will only be given credit once.

Now test yourself

TESTED

1 Write down **three** functions of water. [3 marks]
2 Identify current recommendations for water intake. [1 mark]
3 Explain why fluid is important in the diet of an older person. [2 marks]
4 One of the 'eight tips for eating well' is to drink plenty of water. Explain why this tip is important for children. [3 marks]
5 Adolescents often make their own food choices. Discuss **two** ways an adolescent could make changes to their diet to reduce the chance of dehydration. [4 marks]

9 Nutritional and dietary needs

While all nutrients are needed at each stage, the importance of each one varies depending on factors such as age, gender, body size, level of physical activity, state of health and lifestyle.

Different life stages

School-aged children (4–11 years)

REVISED

- Energy requirements are proportionately high to meet the demands of growth and physical activity.
- Younger children will need small, frequent meals, which are energy dense.
- Older children can eat bulkier meals and the frequency of treats and snacks should be reduced.
- It is important to introduce variety in the diet and to develop healthy eating patterns that can be carried on into adolescence and adulthood.
- It is also important to establish a good dental health routine in childhood, as this will have a positive effect on dental health throughout life.
- The frequency and volume of sugary foods a child eats should be carefully monitored, and drinks should be limited to milk or water to reduce the risk of dental caries.
- Teeth should be brushed with a fluoride toothpaste, after which only water should be consumed.
- To reduce the risk of childhood obesity, children should be encouraged to eat a balanced diet; limit the amount of foods, drinks and snacks high in fat and/or calories; and take regular physical activity.
- Excess salt should be avoided during childhood; there is no need to add salt to children's food.

> **Lifestyle:** an individual's way of living.
>
> **Energy dense:** a food or diet rich in energy.

Table 9.1 **Macronutrient requirements for school-aged children**

Fat	Carbohydrates	Protein
- It is essential that the right type and amount of fat is consumed in childhood to avoid risks to health in later life. - Saturated fats should be replaced with unsaturated fats when possible. - Essential fatty acids play an important role in brain development during childhood.	- Starchy carbohydrates should be the main source of energy during childhood. - Wholegrain varieties should be chosen to reduce the risk of bowel disorders that are common during childhood, such as constipation. - Consumption of sugary carbohydrates should be reduced to lower risks to health including dental caries and childhood obesity.	- Protein requirements increase to facilitate this period of rapid growth. - Protein is required to ensure maintenance and repair of body tissues.

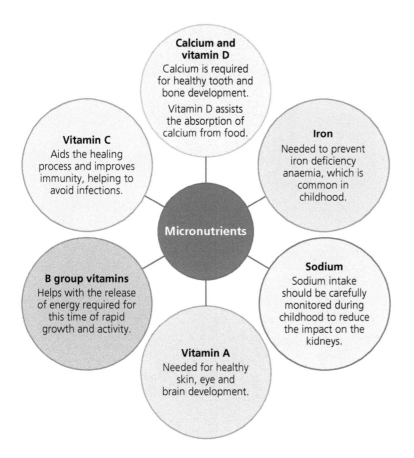

Calcium and vitamin D
Calcium is required for healthy tooth and bone development.
Vitamin D assists the absorption of calcium from food.

Vitamin C
Aids the healing process and improves immunity, helping to avoid infections.

Micronutrients

Iron
Needed to prevent iron deficiency anaemia, which is common in childhood.

Sodium
Sodium intake should be carefully monitored during childhood to reduce the impact on the kidneys.

B group vitamins
Helps with the release of energy required for this time of rapid growth and activity.

Vitamin A
Needed for healthy skin, eye and brain development.

Figure 9.1 **Micronutrient requirements for school-aged children**

Adolescents (12–18 years)

REVISED

- During adolescence, teenagers experience an increase in their rate of growth (height and weight) called the adolescent **growth spurt**.
- This development means that teenagers require more of most nutrients. It is important that food choices are made carefully to support growth and prevent the onset of diet-related disorders that can continue into adulthood.
- Many fast-food options are high in calories, sugar, saturated fat and salt. Regular consumption, combined with inactivity, is a major factor contributing to weight gain during adolescence.
- Breakfast is a very important meal and teenagers should be encouraged to consume a healthy, balanced breakfast that provides 25 per cent of daily energy and nutrients.
- Missing breakfast can lead to **grazing** or snacking on foods high in fat and sugar, an increased risk of dehydration and poor concentration.
- It is vital that adolescents understand the importance of building strong bones as approximately 30 per cent of all minerals deposited in bone throughout life is deposited at this stage.
- **Peak bone mass** is the stage at which bone is at its strongest.
- Achieving good peak bone mass during adolescence can reduce the risk of developing osteoporosis in later life.

Growth spurt: a rapid change in height and weight.

Grazing: food eaten outside normal mealtimes, often mindlessly.

Peak bone mass: the stage at which bone is at its strongest.

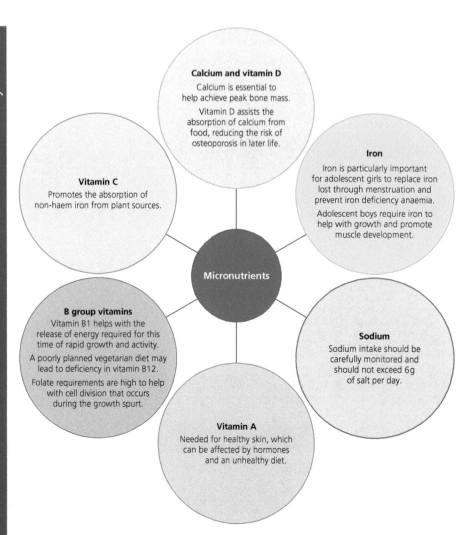

Calcium and vitamin D
Calcium is essential to help achieve peak bone mass.
Vitamin D assists the absorption of calcium from food, reducing the risk of osteoporosis in later life.

Iron
Iron is particularly important for adolescent girls to replace iron lost through menstruation and prevent iron deficiency anaemia.
Adolescent boys require iron to help with growth and promote muscle development.

Vitamin C
Promotes the absorption of non-haem iron from plant sources.

Micronutrients

B group vitamins
Vitamin B1 helps with the release of energy required for this time of rapid growth and activity.
A poorly planned vegetarian diet may lead to deficiency in vitamin B12.
Folate requirements are high to help with cell division that occurs during the growth spurt.

Sodium
Sodium intake should be carefully monitored and should not exceed 6g of salt per day.

Vitamin A
Needed for healthy skin, which can be affected by hormones and an unhealthy diet.

Figure 9.2 **Micronutrient requirements for adolescents**

Table 9.2 **Macronutrient requirements for adolescents**

Fat	Carbohydrates	Protein
• Care is needed to meet increased energy requirements during adolescence to help facilitate growth while also monitoring fat intake.	• Starchy carbohydrates should be eaten, as they are energy dense, rather than sugary carbohydrates that provide empty calories. • The recommended intake of fibre is 25 g per day, which should be maintained to reduce the risk of constipation at this life stage.	• Required to help growth, maintenance and repair of body tissues. • Protein can be used as a secondary source of energy to meet demands of the growth spurt or high levels of physical activity. • Male adolescents have proportionately more muscle than females, and so have higher protein requirements.

Adults, including pregnant women (19–64 years)

REVISED

- From the age of 30, the basal metabolic rate (BMR) begins to slow; intake of high-density, low-nutrient foods such as cakes, biscuits and alcohol should be monitored.

- It is important to maintain a healthy weight by taking regular exercise. Balancing energy intake and energy output reduces the likelihood of excess weight gain and the associated impact on health.

- A way to find out if an adult's weight could have a negative impact on health is to determine their **body mass index (BMI)**. A BMI of over 30 is classified as obese.

> **Body mass index (BMI):** a unit of measurement describing weight in relation to height. BMI is calculated by dividing a person's weight (in kg) by their height squared (m^2). It is used to classify people as underweight, normal weight, overweight or obese.

- As the risk of cardiovascular disease and cancer increase with age, adults are advised to keep up their intake of vitamins A and C by eating at least five portions of fruit and vegetables every day. Vitamin A (carotene) and vitamin C have **antioxidant** properties and form part of the body's defence against dangerous substances called **free radicals**, which are linked to the development of some cancers and cardiovascular disease.
- A diet rich in fibre offers a range of health benefits, including reduced risk of bowel disorders, cardiovascular disease, type 2 diabetes and certain cancers. Adults should therefore include a wide range of fibre-rich foods in their diet.
- As fibre-rich foods are filling and can reduce the need to snack between meals, they can make a positive contribution to weight management.

> **Antioxidant**: a substance that removes potentially damaging substances from the body, e.g. vitamins A and C.
>
> **Free radicals**: dangerous substances linked to the development of some cancers and coronary heart disease.

Table 9.3 **Macronutrient requirements for adults**

Fat	Carbohydrates	Protein
<ul><li>Fat intake needs to be monitored during adulthood, as energy requirements fall when growth stops.</li><li>Too much saturated fat can contribute to obesity, cardiovascular disease, hypertension and diabetes.</li><li>Essential fatty acids omega 3 and omega 6 are important at this life stage due to their positive effect on blood cholesterol and blood pressure.</li></ul>	<ul><li>Carbohydrate requirements should be met by eating complex starchy carbohydrates rather than sugary carbohydrates, which are high in calories but provide a limited range of other nutrients.</li><li>Soluble fibre can help reduce blood cholesterol levels. It also helps control blood sugar levels.</li><li>Insoluble fibre helps to prevent constipation and provides a feeling of fullness, which can decrease the desire to eat between meals and therefore help to maintain a healthy weight.</li><li>The recommended intake of fibre is 30 g per day, which should be maintained to reduce the risk of bowel disorders common at this life stage.</li></ul>	<ul><li>Protein intake should be maintained, as it is required for repair of tissues and recovery from illness.</li></ul>

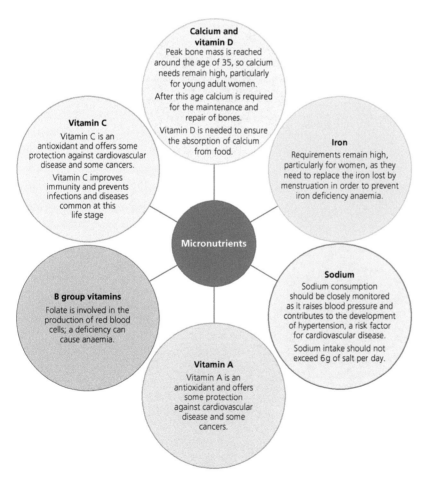

Figure 9.3 **Micronutrient requirements for adults**

Pregnancy

- Pregnant women need to make sure their diet provides enough energy and nutrients for their baby to grow and develop, and for their own body to cope with the changes taking place.

- Pregnancy is divided into three-month periods called trimesters. Nutrient demands increase most rapidly in the third trimester (months 7–9 of pregnancy).

- A pregnant woman needs an extra 200 kilocalories per day to meet her increased energy requirements.

- A new mother's body produces milk for breastfeeding in a process called lactation.

- During lactation a new mother needs to increase her intake of protein, calcium, vitamin D and fluid, to help produce milk.

- Women are advised to eat an extra portion of dairy products during lactation, for example milk on breakfast cereal or yoghurt with fruit as a midday snack.

> **Lactation:** the process of breastfeeding.

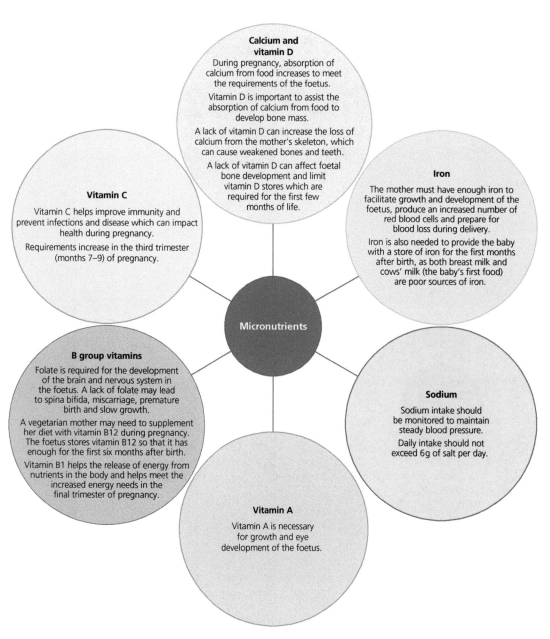

Figure 9.4 **Micronutrient requirements for pregnant women**

- Doctors advise women to take folic acid supplements. When trying to conceive, and in the first 12 weeks of pregnancy, they should take 400 mcg of folic acid per day, in addition to the folate in their diet.
- Taking folic acid supplements reduces the risk of neural tube defects, such as spina bifida, in the foetus.
- Pregnant women are advised to consider taking a supplement of vitamin D (10 mcg) daily to maximise their bone health and that of the foetus.
- Pregnant women are advised not to take vitamin A supplements or multivitamin supplements containing vitamin A, as high doses during pregnancy are linked with a higher risk of birth defects.

Table 9.4 **Macronutrient requirements for pregnant women**

Fat	Carbohydrates	Protein
• Fat intake should be maintained rather than increased during pregnancy. • Essential fatty acids play an important role in reducing high blood pressure in the mother and in the development of the brain/eye in the foetus.	• Increased energy needs should be met by eating starchy carbohydrates, which release energy slowly and provide a feeling of fullness. • Constipation can be a problem during pregnancy, so a diet high in fibre is recommended.	• More protein is needed for growth of the foetus and for repair and maintenance of the mother's body tissue during and after pregnancy.

Foods to avoid in pregnancy

✗ These foods may contain *Listeria* bacteria, which can increase the risk of miscarriage or stillbirth: mould-ripened cheeses, e.g. Brie and Camembert; unpasteurised cheese, e.g. Stilton; pâté; pre-packed salads and 'cook chill' meals.

✗ These foods may contain *Salmonella*, which can cause food poisoning: raw and lightly cooked eggs, and foods that contain them, e.g. mayonnaise and chocolate mousse.

✗ These foods could infect the body with toxoplasmosis, which can cause flu-like symptoms in the mother and damage the nervous system and eyes of the foetus: raw and partly cooked meat, unpasteurised milk and unwashed fruit and vegetables.

✗ These foods contain high levels of vitamin A, which can be toxic if consumed in excess, leading to birth defects: liver and products made from liver, e.g. pâté.

✗ Limit intake of:
  - some types of fish – pregnant women should aim to eat two portions of fish a week but some fish (e.g. swordfish and tuna) should be restricted as they may contain mercury, which can affect the nervous system of the foetus.
  - caffeine – tea, coffee and soft drinks containing caffeine affect the absorption of many nutrients. Pregnant women are advised to have no more than 300 mg of caffeine per day, equivalent to four cups of coffee, six cups of tea or eight cans of cola.
  - alcohol – this can affect the foetus' development. It is particularly damaging if consumed during the first trimester of pregnancy (months 1–3), as it increases the risk of miscarriage, stillbirth, premature labour and foetal alcohol syndrome (a developmental disorder).

Older adults (65 years and over)

REVISED

- Energy requirements decline due to a decrease in the basal metabolic rate and reduced levels of physical activity.
- Dietary guidance from the Eatwell Guide, the 'eight tips for eating well' and lifestyle advice regarding physical activity should still be considered to reduce the risk of health issues that become more common with advancing age, such as cardiovascular disease, dehydration, constipation and osteoporosis.
- Older adults must remember the importance of balancing energy intake and energy use, to achieve a healthy weight. Being overweight or underweight can influence the onset of diet-related diseases such as diabetes and osteoporosis.
- Appetite can decline with advancing age and a varied diet of smaller meals, supplemented with regular snacks, should be eaten.
- The senses of taste and smell may decline, and this can make food seem less appetising. Variety of taste, texture, colour and flavour is the key to sustaining an interest in food.
- Older adults may find that they do not recognise thirst, which means that dehydration may go unnoticed until symptoms have advanced from mild to severe.
- Older adults are more susceptible to dehydration due to changes associated with ageing, including factors such as illness, hormones, kidney function, medication and cognitive function.

Table 9.5 **Macronutrient requirements of older adults**

Fat	Carbohydrates	Protein
• Total fat intake should be monitored to reduce the risk of becoming overweight or obese as estimated average requirements (EARs) decline in older adults. • Saturated fats should be avoided as they increase cholesterol and the risk of cardiovascular disease. • Essential fatty acids should be increased as they help protect against cardiovascular disease.	• Carbohydrate needs should be met by consuming starchy rather than sugary carbohydrates, which are high in calories but provide a limited range of other nutrients. • It is important that older adults still eat a range of foods high in fibre to avoid constipation and other bowel-related disorders common at this stage.	• Protein intake should be maintained to help in the repair of body tissues and recovery from illness.

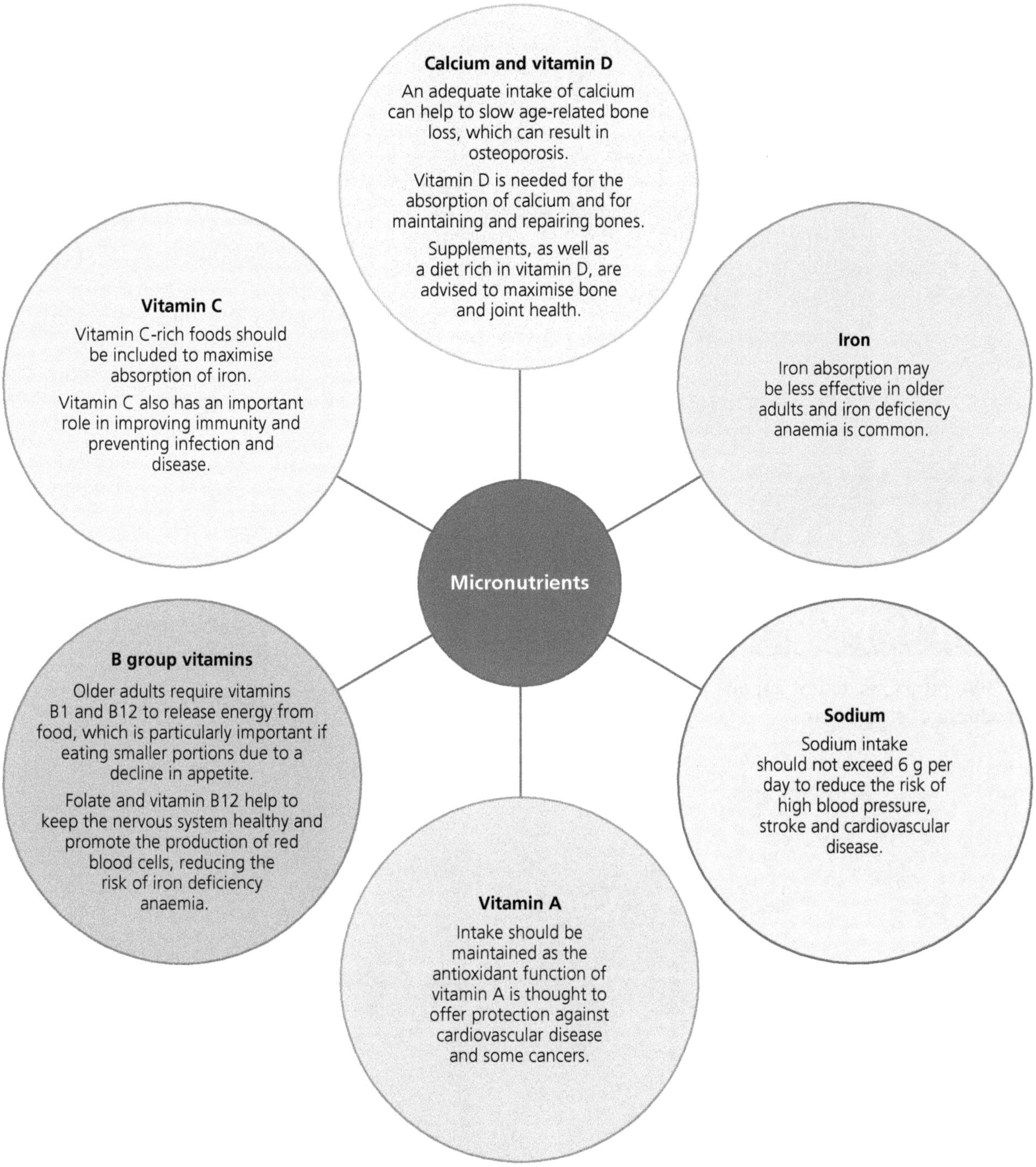

Calcium and vitamin D

An adequate intake of calcium can help to slow age-related bone loss, which can result in osteoporosis.

Vitamin D is needed for the absorption of calcium and for maintaining and repairing bones.

Supplements, as well as a diet rich in vitamin D, are advised to maximise bone and joint health.

Vitamin C

Vitamin C-rich foods should be included to maximise absorption of iron.

Vitamin C also has an important role in improving immunity and preventing infection and disease.

Iron

Iron absorption may be less effective in older adults and iron deficiency anaemia is common.

Micronutrients

B group vitamins

Older adults require vitamins B1 and B12 to release energy from food, which is particularly important if eating smaller portions due to a decline in appetite.

Folate and vitamin B12 help to keep the nervous system healthy and promote the production of red blood cells, reducing the risk of iron deficiency anaemia.

Sodium

Sodium intake should not exceed 6 g per day to reduce the risk of high blood pressure, stroke and cardiovascular disease.

Vitamin A

Intake should be maintained as the antioxidant function of vitamin A is thought to offer protection against cardiovascular disease and some cancers.

Figure 9.5 **Micronutrient requirements of older adults**

Specific lifestyle needs

Vegetarians and vegans

- The Vegetarian Society defines vegetarian as:

  'Someone who lives on a diet of grains, pulses, nuts, seeds, vegetables and fruits with or without the use of dairy products and eggs. A vegetarian does not eat any meat, poultry, game, fish, shellfish or by-products of slaughter.'

- The most commonly adopted vegetarian diets are:
  - **lacto-ovo vegetarian**: eats dairy products and eggs, but does not eat meat
  - **vegan**: does not eat dairy products, eggs, meat or any animal by-products, for example honey or gelatine.

> **Lacto-ovo vegetarian**: a vegetarian who eats dairy products and eggs, but does not eat meat.
>
> **Vegan**: a vegetarian who does not eat dairy products, eggs, meat or any animal by-products, for example honey or gelatine.

Figure 9.6 **Vegetarians do not eat any meat, poultry, game, fish, shellfish or by-products of slaughter**

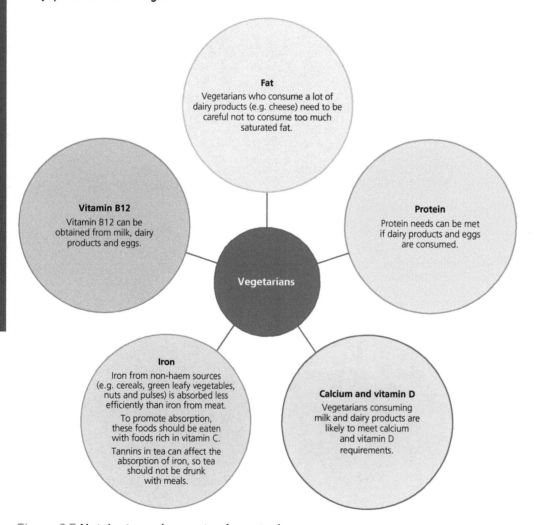

Figure 9.7 **Nutrient requirements of vegetarians**

Quick quizzes at **www.hoddereducation.co.uk/myrevisionnotes**

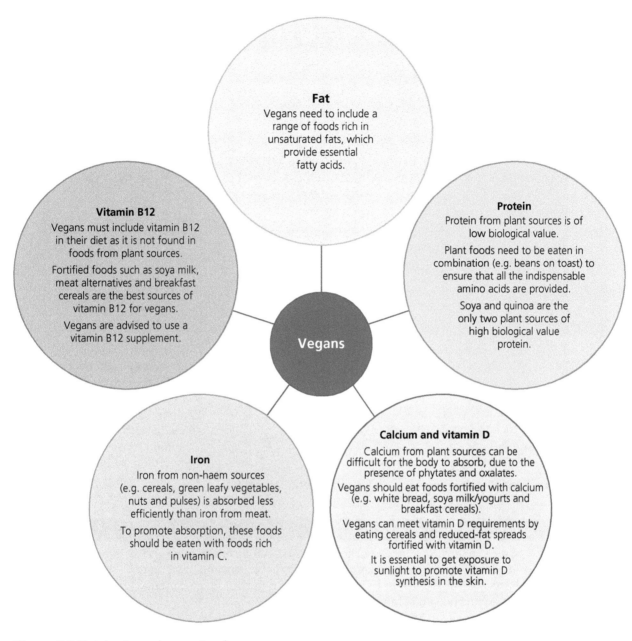

Fat
Vegans need to include a range of foods rich in unsaturated fats, which provide essential fatty acids.

Vitamin B12
Vegans must include vitamin B12 in their diet as it is not found in foods from plant sources.
Fortified foods such as soya milk, meat alternatives and breakfast cereals are the best sources of vitamin B12 for vegans.
Vegans are advised to use a vitamin B12 supplement.

Protein
Protein from plant sources is of low biological value.
Plant foods need to be eaten in combination (e.g. beans on toast) to ensure that all the indispensable amino acids are provided.
Soya and quinoa are the only two plant sources of high biological value protein.

Vegans

Iron
Iron from non-haem sources (e.g. cereals, green leafy vegetables, nuts and pulses) is absorbed less efficiently than iron from meat.
To promote absorption, these foods should be eaten with foods rich in vitamin C.

Calcium and vitamin D
Calcium from plant sources can be difficult for the body to absorb, due to the presence of phytates and oxalates.
Vegans should eat foods fortified with calcium (e.g. white bread, soya milk/yogurts and breakfast cereals).
Vegans can meet vitamin D requirements by eating cereals and reduced-fat spreads fortified with vitamin D.
It is essential to get exposure to sunlight to promote vitamin D synthesis in the skin.

Figure 9.8 **Nutrient requirements of vegans**

Table 9.6 **Dos and don'ts of adopting a vegetarian or vegan diet**

Do	Don't
✔ Eat a variety of foods from the five food groups ✔ Eat fortified cereals for breakfast ✔ Exceed the five-a-day target for fruit and vegetables ✔ Include alternatives to meat to ensure protein needs are met (e.g. soya, quinoa, tofu, beans, lentils, chickpeas, eggs, dairy products, nuts and seeds) ✔ Enjoy a wide variety of food and try new foods (e.g. soya, quinoa, Quorn and pulses)	✘ Give up meat and not replace it with other sources of important nutrients ✘ Depend only on dairy products as a replacement for meat ✘ Eat too many sugary or fatty foods ✘ Rely on ready-made vegetarian meals and convenience foods ✘ Exclude a food group from your diet

Groups with differing energy requirements

People with an active lifestyle that includes sport

- Healthy eating and regular physical activity are important for everyone, but for those with an active lifestyle that includes sport, the quality of their diet and how much they eat and drink can have a dramatic impact on their performance.
- Estimated energy requirements increase substantially for athletes, to meet the body's extra demands during exercise.
- The amount of energy needed depends on the intensity, duration, frequency and type of exercise, and on the age, gender, level of fitness and body fat stores of the individual.
- The World Health Organisation states that in adults, physical activity can include:
  - physical activity during leisure time
  - walking or cycling as a means of transportation
  - work-related physical activity
  - planned exercise such as sport, games or play
  - household chores.

> **Carbohydrate loading**: increasing carbohydrate intake by more than 70 per cent. This is used by those taking part in endurance sports such as marathon running in the three days preceding an event.

Table 9.7 Nutritional and dietary requirements of people with an active lifestyle

Nutrient	Active lifestyle benefit
Carbohydrates	<ul><li>A diet based on complex carbohydrates, such as bread, pasta, rice and beans, is important because they are nutrient-dense, filling and low in fat.</li><li>Carbohydrates are stored in the muscles and liver as glycogen, which is important for endurance and stamina.</li><li>The best way to keep up glycogen levels is to eat a low-fat, high-carbohydrate light meal, two to three hours before exercise; this allows time for digestion and excretion to occur. After exercising, it is important to replenish glycogen stores with low-fat, high-carbohydrate snacks.</li><li>On average, a sportsperson uses 500–1000 kcal per day more than a sedentary adult. Therefore, a small amount of sugary foods and drinks can be consumed. These are not bulky or filling; they provide short bursts of energy quickly and can be an effective way for athletes to top up carbohydrate intake.</li><li>While most sportspeople maintain a high-carbohydrate diet, those involved in endurance events may increase carbohydrate intake by more than 70 per cent in the three days preceding an event. This is known as carbohydrate loading and, combined with a specialist training programme, can increase glycogen stores and maximise performance.</li></ul>
Protein	<ul><li>Most sportspeople can meet their protein requirements by following a balanced diet. Main meals should provide some protein, such as meat, fish, poultry, eggs, dairy products, beans, pulses, Quorn, tofu and nuts.</li><li>Only athletes involved in heavy training, for example weightlifters or marathon runners, need to increase their protein intake in order to gain muscle.</li><li>It is important for all sportspeople to consume enough carbohydrates as fuel. If they do not, their body will use protein as a source of energy, compromising its role in maintaining and repairing tissues, including muscle.</li></ul>

Nutrient	Active lifestyle benefit
Fat	• A high-fat diet makes it difficult to meet carbohydrate requirements and could affect weight management and energy balance.
Vitamins and minerals	• A balanced diet with a variety of foods from the five food groups should provide the vitamins required for good health. • The B group vitamins are particularly relevant for people participating in sport, since they assist with the release of energy from food. • Sportspeople need good iron stores to ensure that oxygen can be transported efficently around the body in their blood and to boost energy stores. • Sportspeople following a vegetarian diet should eat pulses and green leafy vegetables, in combination with taking vitamin C. • Calcium intake should be maintained as, regardless of the sport, everyone should aim to achieve and maintain a good peak bone mass. This contributes to a stronger skeleton, makes bones less susceptible to damage and helps to repair body tissue.
Fluid balance	• Before, during and after an event, athletes must take in fluid to replace what they have lost by sweating. • Water is essential to restore fluid balance in the body, particularly the blood. • People participating in sport need to drink more than the six to eight glasses advised for the general population. • Sports drinks have been developed that claim to offer additional benefits; as they contain sugar and sodium, and are easy to drink so they can increase the rate of rehydration. • Sports drinks are suitable for people involved in vigorous physical activity, but should not be consumed by others, as regular consumption could lead to weight gain and dental caries.

People with a sedentary lifestyle

REVISED

• A person's lifestyle may be described as **sedentary** if it lacks physical activity. However, sedentary behaviour is also where a person spends most of their time sitting or lying down and their energy expenditure is very low.
• Those with a sedentary lifestyle are at greater risk of becoming overweight or obese and need to become more active to achieve energy balance and maintain a healthy weight.

> **Rehydration**: the process of restoring lost water to the body.
>
> **Sedentary**: a lifestyle that is lacking in physical activity; sedentary behaviour is where a person spends most of their time sitting or lying down and their energy expenditure is very low.

Food allergies and food intolerance

Food allergies and food intolerance are both classified as types of food sensitivity.

Food allergies

- An **allergy** to a specific food causes the body's immune system to react.
- Someone with a severe food allergy can experience a life-threatening reaction.
- Individuals with food allergies need to be extremely careful about what they eat.
- The most common symptoms of an allergic reaction include:
  - coughing
  - dry, itchy throat and tongue
  - itchy skin or rash
  - nausea and feeling bloated
  - diarrhoea and/or vomiting
  - wheezing and shortness of breath
  - swelling of the lips and throat
  - runny or blocked nose
  - sore, red and itchy eyes.
- People can deal with nut, egg or fish allergies in the following ways:
  - exclude the food causing the problem from all meals
  - check the labels on all food products
  - identify replacements for foods that cannot be eaten
  - modify recipes and plan meals to meet the specific dietary requirements
  - check menus carefully and ask for advice if unsure when eating out
  - if prescribed, always carry an EpiPen.

> **Food allergy:** an allergy to a specific food that causes the body's immune system to react. Someone with a severe food allergy can experience a life-threatening reaction.

Anaphylaxis

- People with severe allergies can have a reaction called anaphylaxis (pronounced 'anna-fill-axis'), sometimes called **anaphylactic shock**.
- When someone has an anaphylactic reaction, symptoms in different parts of the body may occur at the same time, including rashes, swelling of the lips and throat, difficulty breathing, a rapid fall in blood pressure and loss of consciousness.
- Anaphylaxis can be fatal if it is not treated immediately, usually with an injection of adrenaline (epinephrine). This is why it is extremely important for someone with a severe allergy to take their medication with them wherever they go.

> **Anaphylactic shock:** a sudden, severe and potentially fatal allergic reaction. Symptoms may include a drop in blood pressure, difficulty breathing, itching and swelling.

Food intolerance

- **Food intolerance** does not involve the immune system and is generally not life-threatening.
- If someone eats a food they are intolerant to it could make them feel ill or affect their long-term health.
- The most common symptoms of intolerance to a particular food include:
  - stomach cramps
  - bloating
  - nausea
  - diarrhoea
  - tiredness.

> **Food intolerance:** if someone eats a food they are intolerant to it could make them feel ill or affect their long-term health. It does not involve the immune system as is generally not life-threatening.

Lactose intolerance

Table 9.8 **Lactose intolerance**

Definition	Symptoms	Foods to avoid	Dietary advice
• Lactose is a sugar found naturally in milk. In the body, it is broken down by an enzyme called lactase and absorbed into the bloodstream. • Lactose intolerance is caused by a shortage of lactase. • When someone does not have enough of this enzyme, lactose is not absorbed properly from the gut.	• The main symptoms are stomach cramps, bloating, flatulence and diarrhoea.	• Doctors advise people with lactose intolerance to avoid consuming cows' milk.	• Some products made from cows' milk can be eaten, including hard cheeses such as Cheddar, which contain little lactose, and yoghurt, since the live bacteria it contains promotes the digestion of lactose. • Some people with lactose intolerance use soya milk as an alternative to cows' milk. • It is important to take advice from a doctor before making any changes to the diet that result in a food group being excluded.

Allergy labelling

Under EU labelling legislation, if any of the following ingredients are added to pre-packed foods or drinks in the UK, they must be indicated clearly on the label:

- celery
- cereals that contain gluten
- crustaceans
- eggs
- fish
- lupin
- milk
- molluscs
- mustard
- tree nuts (including almonds, hazelnuts, walnuts, brazil nuts, cashews, pecans, pistachios and macadamia nuts)
- peanuts
- sesame seeds
- soya
- sulphur dioxide.

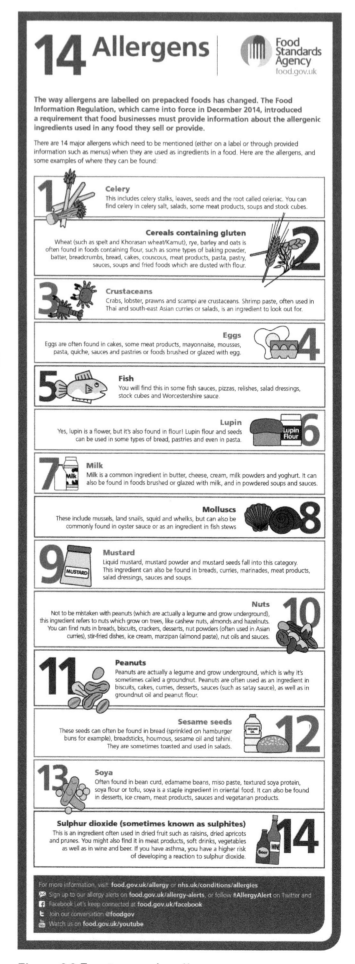

Figure 9.9 **Fourteen major allergens**

Calculating energy and nutritional values of recipes, meals and diets

- The amount of energy in food is measured in kilocalories (kcal):
  - 1 gram of fat provides 9 kilocalories
  - 1 gram of carbohydrate provides 3.75 kilocalories
  - 1 gram of protein provides 4 kilocalories.
- To calculate the energy in a specific food, multiply the number of grams of that food by the energy value in kilocalories for each nutrient that is in the food.

Example

100 g of Greek yoghurt contains 0.4 g of fat, 3.6 g of carbohydrate and 10 g of protein.

Total energy from fat = 0.4 × 9 = 3.6

Total energy from carbohydrate = 3.6 × 3.75 = 13.5

Total energy from protein = 10 × 4 = 40

Total energy from 100 g Greek yoghurt = 57.1 kcal

Total energy from 1 g Greek yoghurt = 0.571 kcal

You can use this information to calculate how much energy is in a recipe. For example, if a recipe uses 50 g of Greek yoghurt, multiply 0.571 × 50 = 28.55. This is the energy supplied by Greek yoghurt in this recipe.

The energy value of individual ingredients in a recipe, meal or diet can be calculated in this way.

Now test yourself

TESTED ☐

1. Which of the following foods would not be eaten by a vegan?
   Choose **two** from: [2 marks]
   - bread
   - lentils
   - pasta
   - custard
   - brocolli
   - jelly.
2. Explain why an adolescent should have a good intake of calcium. [3 marks]
3. Explain the difference between a food allergy and food intolerance. [4 marks]
4. Explain why the following nutrients are important in the diet of an older adult: [6 marks]
   (a) protein
   (b) carbohydrates.
5. Women are advised to avoid certain foods when pregnant. Analyse this statement and give specific example of foods they should avoid. [9 marks]

Exam tip

Ensure that you are able to give specific nutritional information related to the needs of each of the different groups of people. Avoid just explaining the nutrients in general terms. For example: 'Calcium is essential in adolescence to help achieve peak bone mass; however, in older adults calcium helps to slow age-related bone loss.'

Typical mistake

Responses can often be written in general terms. For example, 'protein is required for growth and repair'. This is not accurate when answering a question focused on older adults; protein is not required for growth at this stage, but it is needed for the maintenance of body tissue.

10 Priority health issues

A number of dietary and lifestyle factors contribute to the development of a range of health issues.

Obesity

REVISED

- The World Health Organisation (WHO) defines overweight and **obesity** as 'abnormal or excessive fat accumulation that may impair health'.
- Body mass index (BMI) is used by medical practitioners to determine overweight and obesity in adults.
- WHO defines overweight as a BMI equal to or more than 25, and obesity as a BMI equal to or more than 30. This method of classifying body weight applies to adult men and women, except during pregnancy and older age. Children have their own BMI range, which takes age and gender, as well as height and weight, into account.
- Fat distribution is also a useful guide to determine overweight and obesity e.g. waist measurement.
- Being obese is a major **risk factor** for a number of other diseases, including cardiovascular disease, **hypertension**, type 2 diabetes, joint disorders such as osteoarthritis, and some cancers.
- Obesity is becoming more common among school-aged children and pregnant women. Childhood obesity is a major health issue, as overweight children are more likely to become obese adults.
- To achieve and maintain a healthy weight, a balanced diet and regular physical activity are essential.

> **Obesity**: a dietary disorder caused when excess kilocalories are stored as fat, resulting in excessive weight gain. Obesity is classified as a BMI over 30.
>
> **Risk factor**: something that increases the likelihood of developing a disease.
>
> **Hypertension**: high blood pressure.

Table 10.1 Dietary and lifestyle factors and advice related to obesity

Dietary and lifestyle factors that may contribute to the development of obesity	Dietary and lifestyle advice to manage obesity
<ul><li>Increased consumption of energy-dense food high in calories, fat and sugar</li><li>Decreased levels of physical activity</li><li>Overconsumption of fast food, processed food and treat foods</li><li>Larger portion sizes</li><li>Increased snacking and grazing</li><li>High intake of sugary drinks</li><li>Comfort eating</li><li>Lack of physical activity</li><li>Sleep deprivation</li><li>Genetic traits such as a large appetite</li><li>Medical conditions linked to hormones</li><li>Certain medications, including steroids, which can increase appetite and, if taken long term, can contribute towards weight gain</li></ul>	<ul><li>Adult men should not exceed 2500 kilocalories per day and adult women should not exceed 2000 kilocalories</li><li>Eat more fruit, vegetables, pulses and wholegrains</li><li>Choose lower-fat, lower-sugar dairy options</li><li>Drink six to eight glasses of fluid a day</li><li>Check food labels carefully</li><li>Eat treat foods less often and in small amounts</li><li>Children should engage in 60 minutes of moderate intensity physical activity a day</li><li>Adults should aim for approximately 150 minutes of physical activity per week</li><li>Most healthy adults need 7.5–9 hours of sleep per night</li></ul>

Cardiovascular disease

REVISED

- **Cardiovascular disease** is a general term for conditions affecting the heart or blood vessels.
- The coronary arteries supply blood to the heart. Heart disease can occur when one or more of the coronary arteries is narrowed or completely blocked by a build-up of fatty deposits called **cholesterol** on its walls; this condition is called **atherosclerosis**.
- When the arteries become partly or totally blocked, the heart has to work much harder to pump blood and oxygen around the body.
- If the arteries become totally blocked, blood cannot flow to the heart and this can result in a **heart attack**.
- Symptoms of a heart attack include pain in the centre of the chest, the arms, back, neck, jaw or stomach; shortness of breath; breaking out in a cold sweat and nausea.
- Cardiovascular disease can often be prevented and there are many things that people can do to protect their heart and keep it healthy, whatever their age.

> **Cardiovascular disease:** general term for conditions affecting the heart or blood vessels.
>
> **Cholesterol:** a fatty substance found in the blood. It is mainly made in the body and plays an essential role in how every cell works. Too much cholesterol in the blood can increase the risk of heart problems.
>
> **Atherosclerosis:** a condition in which an artery wall thickens as a result of the build-up of fatty materials such as cholesterol.
>
> **Heart attack:** a sudden illness caused when the arteries supplying blood and oxygen to the heart become blocked.

Healthy artery

Normal blood flow

Blocked artery

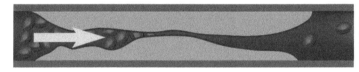

Fatty deposits block artery, obstructing the blood flow

Figure 10.1 This cross-section drawing of a coronary artery shows how fatty deposits can build on the artery walls and narrow the inside of the artery

Table 10.2 Dietary and lifestyle factors and advice related to cardiovascular disease

Dietary and lifestyle factors that may contribute to the development of cardiovascular disease	Dietary and lifestyle advice to manage cardiovascular disease
<ul><li>High levels of cholesterol caused by eating too much saturated fat</li><li>Eating too much salt</li><li>Too little fibre, which can increase blood cholesterol</li><li>A diet low in antioxidants</li><li>Being overweight</li><li>Too much alcohol, which can damage the heart muscle</li><li>Lack of physical activity</li><li>Smoking</li></ul>	<ul><li>Saturated fats should be replaced with monounsaturated and polyunsaturated fats</li><li>The essential fatty acids omega 3 and omega 6 are particularly important</li><li>Adults should not exceed 6 g of salt per day</li><li>Soluble fibre from grains, pulses and fruit can help lower blood cholesterol levels</li><li>Fruit and vegetables are a valuable source of the antioxidant nutrients vitamin A (beta carotene) and vitamin C, while wholegrain cereals are a good source of vitamin E</li><li>Maintain a healthy weight</li><li>Limit intake of alcohol</li><li>Be physically active</li><li>Avoid smoking</li></ul>

Type 2 diabetes

- **Diabetes** is a condition in which the amount of **glucose** in the blood is too high.
- Diabetes develops either because the body does not produce **insulin** as needed, or because the insulin that it does produce does not work effectively.
- There are two types of diabetes:
  - In type 1 diabetes, the body cannot make any insulin. Type 1 diabetes occurs more commonly in children, adolescents and young adults. It is also known as insulin-dependent diabetes.
  - In type 2 diabetes, not enough insulin is produced, or the insulin produced in the body does not work effectively. Type 2 diabetes tends to develop gradually as people get older – usually after the age of 40. It is also known as non-insulin dependent diabetes.
- The cause of type 1 diabetes is uncertain. It is an auto-immune condition.
- The risk factors for type 2 diabetes are much clearer.

> **Diabetes:** a condition in which the body's normal way of breaking down sugar is not functioning properly. This means that the pancreas is not producing any or enough insulin to regulate the amount of sugar in the blood.
>
> **Glucose:** a sugar naturally found in honey and the juices of many fruits. Glucose is also the sugar that circulates in the blood.
>
> **Insulin:** a hormone that the body needs to convert sugar into energy.
>
> **Stroke:** an attack caused by the blockage of an artery carrying blood to the brain.

Table 10.3 **Dietary and lifestyle factors and advice related to type 2 diabetes**

Dietary and lifestyle factors that may contribute to the development of type 2 diabetes	Dietary and lifestyle advice to manage type 2 diabetes
<ul><li>The risk of developing type 2 diabetes increases with age, so the older you get, the more at risk you are</li><li>The risk of developing type 2 diabetes increases if you are overweight or if your waist measurement is:<ul><li>31.5 inches or over for a woman</li><li>37 inches or over for men</li></ul></li><li>Being inactive can contribute to weight gain and increase your risk of developing type 2 diabetes</li><li>The risk of developing type 2 diabetes increases if diabetes exists in the family. The closer the relative is – for example, mother, father, brother or sister – the greater the risk</li><li>Other health problems, such as being diagnosed with circulation problems, having a heart attack, **stroke** or high blood pressure, lead to an increased risk of diabetes</li></ul>	<ul><li>Eat a healthy, balanced diet, low in fat, sugar and salt</li><li>Include plenty of fruit and vegetables</li><li>Eat a wide variety of foods as part of a healthy diet</li><li>Weight loss</li><li>Increased physical activity</li></ul>

Osteoporosis

- **Osteoporosis** is a disease characterised by low bone density and deterioration of the bone tissue, which leads to fragile bones and an increased risk of fractures.
- The condition occurs through the natural process of ageing. From around the age of 35 more bone cells are lost than replaced, which results in a decrease in bone density.
- Broken wrists, hips and spinal bones are the most common fractures in people with osteoporosis, however fractures can occur in any bone.
- Osteoporosis affects all age groups, but it is most common in post-menopausal women.
- **Peak bone mass** is reached at the age of 30–35 years and is the stage at which the bone is strongest. After this age, bone mass decreases, as more bone cells are lost than made.
- Achieving good peak bone mass during adolescence and early adulthood means that bones are stronger before natural age-related bone loss begins. This can reduce the risk of developing osteoporosis in later life.

> **Osteoporosis:** a disease characterised by low bone density and deterioration of the bone tissue, which results in fragile bones and increased risk of fractures.
>
> **Peak bone mass:** the stage at which bone is strongest, reached at the age of 30–35 years.

Figure 10.2 **Developing osteoporosis: a strong dense bone (left) and a fragile osteoporotic bone (right)**

Table 10.4 **Dietary and lifestyle factors and advice related to osteoporosis**

Dietary and lifestyle factors that may contribute to the development of osteoporosis	Dietary and lifestyle advice to manage osteoporosis
<ul><li>Insufficient intake of calcium, vitamin D and protein when bones are growing and developing may affect peak bone mass</li><li>Being underweight or overweight can increase the risk of fractures</li><li>Alcohol consumption</li><li>Smoking</li></ul>	<ul><li>Eat a healthy diet, including foods rich in calcium, vitamin D and protein</li><li>Vitamin D is made in the skin when it is exposed to sunlight</li><li>Maintain protein intake</li><li>Maintain a healthy body weight</li><li>Weight-bearing exercise, such as jogging, aerobics, tennis, skipping, dancing and brisk walking, is good for strengthening bones</li><li>Reduce alcohol intake</li><li>Stop smoking</li></ul>

Dental caries

REVISED

- Dental caries (tooth decay) occur when plaque acids attack and soften the enamel and dentine of the tooth.
- Plaque is a thin, sticky film on the surface of teeth. It contains many different types of bacteria, which react with free sugars in food and drinks, creating acid.
- This acid dissolves minerals (e.g. calcium) from the enamel in a process called demineralisation. Snacking or grazing has an impact on tooth decay because it increases the time teeth are exposed to acid and increases the frequency of demineralisation.

> Dental caries: tooth decay.
>
> Enamel: the thin, hard outer covering of the tooth.
>
> Dentine: tissue that lies under the enamel of teeth.
>
> Demineralisation: when acid dissolves minerals from tooth enamel.
>
> Remineralisation: when minerals are restored to tooth enamel.

 + 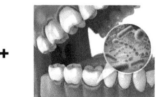 = **ACID**

Food and drinks containing free sugars   **bacteria in plaque**   **acid**

ACID +  =

Acid   **healthy tooth**   **dental caries**

Figure 10.3 **How dental caries occur**

- Acid attacks can last for over an hour after eating or drinking.
- Saliva plays an important role in dental health, neutralising the acid and returning calcium to tooth enamel.
- Children and adolescents are particularly at risk of dental caries.
- Young children are more at risk because the enamel on their teeth, which have just emerged, is not very strong and so is susceptible to acid attack.

Table 10.5 **Dietary and lifestyle factors and advice related to dental caries**

Factors that may contribute to the development of dental caries	Advice to manage dental caries
<ul><li>A diet high in foods and drinks containing free sugars</li><li>Overconsumption of acidic fruit juices and fizzy drinks</li><li>Regular snacking and grazing</li><li>Poor dental hygiene</li><li>Diet low in fibre and calcium</li></ul>	<ul><li>Eating nutritious meals and avoiding snacking in between</li><li>Some foods, such as milk and dairy products, may protect against dental caries, since they appear to reduce acid in the mouth</li><li>Drinking water increases the flow of saliva, which neutralises the acid produced by plaque</li><li>Eating foods rich in fibre stimulates the production of saliva, which neutralises acid in the mouth and encourages remineralisation</li><li>Fluoride has a role to play in protecting teeth against decay</li><li>Visiting the dentist regularly – preferably every six months</li><li>Good oral hygiene includes making sure that plaque is removed from teeth effectively by brushing correctly and using dental floss</li></ul>

Iron deficiency anaemia

- Iron is required for the formation of **haemoglobin**, which is responsible for transporting oxygen around the body in red blood cells. A lack of iron in the blood can lead to **iron deficiency anaemia**.
- Symptoms of iron deficiency anaemia include weakness, faintness, dizziness, lethargy and sometimes headaches, palpitations and sore gums. Someone who is anaemic may look pale.
- Anaemia most commonly affects growing children, adolescents (particularly girls), women and older adults.
- Pregnant women must watch out for anaemia, as pregnancy is a time when the body may require additional iron.
- Vegetarians must plan their diet carefully to ensure that they get enough iron from non-meat sources.

> **Exam tip**
>
> Read the question carefully and respond in context to the specific health issue.

> **Haemoglobin:** the red oxygen-carrying pigment in red blood cells.
>
> **Iron deficiency anaemia:** a condition caused by a lack of iron, meaning that red blood cells cannot carry enough oxygen around the body.
>
> **Menstruation:** vaginal bleeding that occurs as part of a woman's monthly period.

Table 10.6 **Dietary and lifestyle factors and advice related to iron deficiency anaemia**

Factors that may contribute to the development of iron deficiency anaemia	Advice to manage iron deficiency anaemia
• Haem iron from animal sources (e.g. liver, meat and fish) is readily absorbed by the body; not eating enough of these foods can be a cause of iron deficiency anaemia • Non-haem iron from green leafy vegetables, pulses and fortified cereals is less readily absorbed by the body • Some chemicals found naturally in food, such as phytates in certain fibre-rich foods and tannins in tea, can interfere with how much iron is absorbed • Anaemia is more likely to develop during pregnancy if the mother's diet is low in iron • Introducing cows' milk into a baby's diet before the age of 12 months can lead to iron-deficiency anaemia because cows' milk is low in iron • Boys are at risk of developing iron deficiency anaemia during the first stages of puberty, due to rapid growth • Girls are at risk due to **menstrual** blood loss and smaller iron stores • Anaemia is common in women of all ages who have heavy periods	• Eat a wide range of foods rich in haem iron, including red meat, offal and fish • Eat a wide range of foods rich in non-haem iron, including green leafy vegetables (e.g. spinach, kale and watercress), pulses (e.g. lentils) and dried fruit (e.g. apricots) • Eat plenty of foods rich in vitamin C (e.g. peppers, broccoli, Brussels sprouts, oranges and kiwi fruit); vitamin C helps the body to absorb iron from non-haem sources • Infants under 12 months should not be given cows' milk as a main drink; breast milk or specially fortified infant formula milk are the only suitable choices • Iron-rich foods should be introduced during weaning after six months of age • Ensure that vegan or vegetarian diets include a good range of iron-rich foods • Do not take iron supplements unless advised to do so by a doctor. High doses of iron can be dangerous, especially in young children

Now test yourself

1 Describe type 2 diabetes. [2 marks]
2 Write down **three** good sources of non-haem iron. [3 marks]
3 Give **two** reasons why obesity is increasing in children. [4 marks]
4 Discuss **three** pieces of diet and lifestyle advice to reduce dental caries. [6 marks]
5 Discuss diet and lifestyle advice for an individual wishing to reduce their risk of cardiovascular disease. [9 marks]

> **Typical mistake**
>
> Candidates often do not differentiate between dietary and lifestyle factors. Some questions ask for dietary advice only, some for lifestyle advice, and others ask for both.

11 Being an effective consumer when shopping for food

A **consumer** is anyone who buys a product, or uses a service, in either the **public sector** or the **private sector**.

What is an effective consumer? REVISED

Being an effective consumer includes:

- knowing about your consumer rights and responsibilities.
- being aware of where to find expert consumer advice.
- being able to deal confidently with issues such as complaining about faulty goods or poor service and making your voice heard.

Barriers to being an effective consumer REVISED

- There are a range of barriers to being an effective consumer.

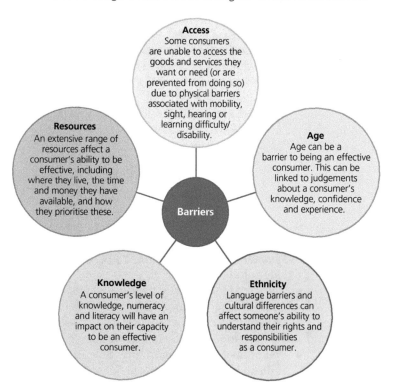

Access
Some consumers are unable to access the goods and services they want or need (or are prevented from doing so) due to physical barriers associated with mobility, sight, hearing or learning difficulty/disability.

Resources
An extensive range of resources affect a consumer's ability to be effective, including where they live, the time and money they have available, and how they prioritise these.

Age
Age can be a barrier to being an effective consumer. This can be linked to judgements about a consumer's knowledge, confidence and experience.

Barriers

Knowledge
A consumer's level of knowledge, numeracy and literacy will have an impact on their capacity to be an effective consumer.

Ethnicity
Language barriers and cultural differences can affect someone's ability to understand their rights and responsibilities as a consumer.

Figure 11.1 **Barriers to being an effective consumer**

> **Consumer:** anyone who buys a product or uses a service.
>
> **Public sector:** the part of the economy that is owned and controlled by the government.
>
> **Private sector:** the part of the economy that is owned and controlled by private individuals and business organisations.
>
> **Economies of scale:** a decrease in the cost per unit of a product when larger quantities are produced.

Food shopping options REVISED

- There are many different food shopping options available, and each has advantages and disadvantages for different consumers.

Table 11.1 **Evaluating food shopping options for a range of consumers**

Food shopping option	Advantages	Disadvantages
Independent grocery shops	• Offer personal and friendly service • Usually in residential areas or town centres for ease of access • Often sell things in small quantities, helping customers to save money and reduce waste • Many offer artisan, locally produced products	• May be more expensive • Opening hours may be restricted • The range of products for sale may be limited • Access and parking may be restricted
Supermarkets	• Economies of scale means they can offer value for money • Offer an extensive range of products, services and facilities • Offer a range of financial incentives (e.g. loyalty cards) • Many have extended opening hours (e.g. 24-hour opening)	• Usually situated out of town, therefore transport is essential • Often busy and large, making shopping time consuming • Special offers and the extensive range of food available can encourage impulse buying, which increases expenditure and may contribute to food waste • May be a lack of local produce available
Markets	• May be cheaper than shops, particularly when buying in small amounts • Local and seasonal produce are promoted and widely available • Many markets offer an extensive range of food products • Friendly, sociable shopping experience	• Packaging and labelling may not be available to help determine quality • Outdoor markets are dependent on favourable weather conditions • Usually only operate on specific days and early in the morning • Access can be an issue, e.g. parking in busy town centres where food markets are often located
Farm shops	• Consumers can support local farmers and small businesses • It can be more sustainable to eat locally grown food (e.g. reduces food miles) • Friendly, sociable shopping experience as it is usually a family-run business • Consumers can access artisan products they may not find elsewhere (e.g. local honey)	• May be more expensive, as production methods may be more costly • Limited range of special offers and price promotions when compared to other shopping options • Limited opening hours • Need access to car as often situated in rural locations
Online shopping	• Available 24 hours a day, 365 days a year • Extensive range of products and services can be purchased, including specialist and overseas brands • Price checking a wide range of retailers can save money • Click and collect or home delivery schemes make shopping online convenient	• Online shoppers cannot personally examine and select items and this may affect the quality of food received • Consumers must have a credit or debit card to buy online, which may raise security issues • Goods and services can be time consuming to compare and may be misrepresented • Delivery/return charges can be expensive, adding to the overall cost of products
Shopping apps	• Food shopping apps make ordering quick and easy • Some allow you to order from an extensive range of takeaway menus (e.g. Just Eat) • Some deliver meals from nearby food retailers that do not normally offer a takeaway service (e.g. Deliveroo) • There are a number of payment options, e.g. cash on delivery, PayPal, Apple Pay or by debit or credit card	• Could encourage over-reliance on fast food, which could have a negative impact on health in the long term • May discourage food preparation at home, which could have an impact on food-handling skills • Can be an expensive way to buy food • Consumers need to have access to the apps and feel confident about using them

The Northern Ireland Trading Standards Service

REVISED

- The role of the Northern Ireland Trading Standards Service is to:
  - O promote and maintain fair trading
  - O protect consumers
  - O enable reputable businesses to thrive within Northern Ireland
  - O enforce a wide range of consumer protection laws
  - O provide advice and guidance to both consumers and traders.
- The Northern Ireland Trading Standards Service deals with the following areas related to food:

1

Weights and measures	All food sold by weight or volume must be accurately measured.

2

Pricing	It is illegal for retailers to give consumers price indications that are misleading, or give price comparisons that are not fair or meaningful. Food retailers must display the price of all goods they offer for sale and must also show the unit price of goods that are pre-packed in fixed quantities.

3

False or misleading descriptions	Food or food-related services must be accurately described when offered for sale.

Figure 11.2 **Areas covered by the Northern Ireland Trading Standards Service**

Exam tip

You may be asked to evaluate shopping options for a specific person or group of people. Ensure that you include a range of advantages and disadvantages to demonstrate the 'evaluate' skill.

Typical mistake

Responses to questions on barriers to being an effective consumer can often be general and lacking focus on the question being asked. There is often a misunderstanding of the following terms: access, age, ethnicity, knowledge and resources.

Now test yourself

TESTED ☐

1 Write down the meaning of the term 'effective consumer'. [2 marks]

2 Describe **two** barriers that may prevent an individual with a physical disability from being an effective consumer. [4 marks]

3 Explain **three** ways the Northern Ireland Trading Standards Service protects consumers when shopping for food. [6 marks]

4 Discuss **two** advantages and **one** disadvantage for an older adult shopping in a local market. [6 marks]

5 Evaluate online shopping as a way of buying food. [9 marks]

12 Factors affecting food choice

The choices people make about what food to buy are influenced by various factors.

Factors affecting individual food choice

REVISED

Figure 12.1 **Factors affecting individual food choice**

Personal, social and economic

- Personal characteristics and circumstances, such as age, personality, likes and dislikes, occupation, lifestyle and the amount of free time available, influence a consumer's food shopping decisions. These **personal factors** tend to change at different life stages.
- **Social factors** are those relating to the influence that other people have on us. For example, we are influenced by our family, friends and any groups with which we identify, such as church groups or sports teams. Many shopping decisions are a mixture of social interaction and individual decision making.
- **Economic factors** have a major influence on food shopping. During periods of economic growth, people shop more frequently and often buy on impulse. It is very different during periods of economic decline. Consumers with more disposable income have a wider range of foods to choose from and may choose to eat out more often.

> **Personal factors**: factors relating to an individual.
>
> **Social factors**: factors relating to the influence of others.
>
> **Economic factors**: factors relating to financial issues.

Religious and cultural

- **Religious factors** can affect both the food someone chooses to buy and how that food is prepared. Religious beliefs can influence the food that people can eat or must avoid.
- **Cultural factors** can determine what a particular group eats, how they eat, where they cook and the opportunities they have to travel and experience different food cultures.

Figure 12.2 **In Islam, permissible foods and food preparation methods are classified as halal**

Ethical and environmental

- **Ethical factors** when purchasing food involve thinking about the welfare of the people, animals and communities involved in producing or providing food.
- **Environmental factors** are something that everyone has a responsibility to consider when choosing food. This includes the energy used in food production processes, food packaging, recycling, organic food and food miles.

Health

- Health concerns may cause people to exclude foods from their diet for medical reasons or to prevent disease. The influence of **health issues** on food shopping depends on consumers' interest in their own health and well-being.

Religious factors: factors relating to a person's faith.

Cultural factors: factors relating to customs and traditions in particular societies.

Ethical factors: factors relating to morals.

Environmental factors: factors relating to global resources.

Health issues: factors relating to well-being.

Figure 12.3 **Someone with an allergy or intolerance may need to exclude certain foods from their diet**

Marketing strategies that influence consumer food choice

- Marketing is a complex process that ultimately aims to persuade consumers to buy a particular product or service. Each day we are bombarded with an extensive range of **marketing strategies**, which companies hope will encourage us to spend money.

Financial incentives

- **Financial incentives** are promotions offered by food retailers that are of economic benefit to consumers. They are an important element of a food business' marketing strategy because they increase profitability.
- Price promotions may take the form of a discount to the normal selling price of a product, for example 'buy one get one free' or '20% extra free'.
- Money-off vouchers are often used to offer a current or future discount to consumers.
- Loyalty cards offered by some food retailers allow consumers to collect points that can be exchanged for money-off vouchers or other rewards, for example restaurant vouchers.
- Larger stores offer their own brands as cheaper alternatives to branded products.
- Price checking is when a retailer highlights products that they are offering at lower prices than other stores, to show that they are giving better value for money.

Store layout

- Every food retailer uses strategies regarding their **layout** to encourage consumers through the door and to get them to stay, because the longer they stay, the more they spend.
- At the entrance, just inside the supermarket, there is some clear space to let customers adjust to the atmosphere. The heating may also blow warm air on to customers, which is psychologically welcoming.
- Magazines or seasonal offers may be displayed in this area, often to the right of the front door. This encourages people to stay and browse, before they do the rest of their shopping.
- Fruit and vegetables are placed near the front of supermarkets. Consumers associate fruit and vegetables with freshness and quality, and having them at the front has a positive effect on sales.
- Bread and milk are essential purchases for most customers. They are usually displayed at the back of supermarkets so that customers going to buy them have to walk past displays of other goods that might appeal to them.
- Displays at the checkout are the supermarket's last chance to tempt customers to buy more. Checkout displays may include sweets, but many other goods are put here too, often depending on the weather or the season.

> **Marketing strategies:** methods used by retailers to persuade consumers to spend money.
>
> **Financial incentives:** promotions offered by food retailers that are of economic benefit to consumers.
>
> **Store layout:** the floor plan of a store and the placement of products, which can influence consumer behaviour and spending.

Figure 12.4 'Eye level is buy level' is one strategy food retailers use to promote certain products

Advertising

- One marketing strategy used to promote products and services is **advertising**. The main purpose of advertising is to:
  - ○ reach prospective customers and influence their buying behaviour
  - ○ develop increased levels of brand awareness
  - ○ communicate the qualities and benefits of specific foods
  - ○ remind consumers about new or existing foods
  - ○ generate repeat purchases.
- Advertisers carefully select images, language, music and celebrities to make their products more appealing to their target consumers.
- Advertisements often provide limited factual information and instead cater to consumers' desires, dreams and aspirations.
- Age is widely used for targeting purposes.
- Packaging is an important element of food advertising. Bright, colourful, glossy packaging creates a perception of quality and increases the likelihood of purchase.

> **Advertising**: messages that are intended to inform or encourage particular consumer behaviour and spending.

Table 12.1 **Advertising examples**

Contexts	Examples
Media	Newspapers, magazines, television and cinema
Outdoors	Billboard posters, transportation (e.g. bus shelters), illuminated signs
Direct mail	Leaflets, flyers and vouchers
Sponsorship	Sports events, entertainment events and charitable events
Online	Email, pop-ups, web banners, mobile advertising, social media

Mandatory and voluntary information on food labels and packaging

Food labels serve three main purposes:

- to provide information about the food/drink
- to distinguish the food/drink from others available
- to provide information so consumers can decide whether the food/drink is safe to eat.

Mandatory information on food labels and packaging

- In Northern Ireland, food labelling is controlled by the Food Information Regulations (Northern Ireland) 2014.
- Environmental Health Officers within local district councils enforce these regulations in Northern Ireland.
- Information provided on a food label should be accurate and not misleading.
- It is **mandatory** that the information in Table 12.2 appears on a food/drink label.

> **Mandatory:** required by law.
>
> **Use by date:** the date by which food must be eaten; after this date it is likely to become unsafe to eat and could cause food poisoning.
>
> **Best before date:** the date before which food will be at its best. Foods will remain safe to eat after this date, but the quality may be affected.
>
> **Date marks:** date labels found on food products (e.g. use by and best before).

Table 12.2 Mandatory information that must appear on food/drink labels

1	Name of the food	The name must inform consumers of the precise nature of the product, e.g. goat's cheese and tomato pizza.
2	List of ingredients	Ingredients used in the product must be listed in descending order. This must be provided with a heading that includes the word 'ingredients'.
3	Quantity of certain ingredients	The quantities of any ingredients that are emphasised on a label to categorise a food should be shown. This is called the Quantitative Ingredient Declaration (QUID). The minimum percentage of the ingredient must be in the ingredients list or beside the name of the food.
4	Net quantity	The actual weight of the food, not including the packaging, must be stated in metric units (kilograms, grams, litres and millilitres).
5	Indication of minimum durability (for example, 'use by' or 'best before' dates)	Two main types of date marks are required: • Best before: these are put on foods that would be expected to be fit to eat and retain their quality for more than 18 months. 'Best before' dates are more about quality than safety. After its 'best before' date, the food does not become harmful, but it might begin to lose its flavour and texture. • Use by: food manufacturers must state a 'use by' date on products that are highly perishable or go off quickly. This is the date by which the food should be eaten. Food or drink should not be consumed after the end of the 'use by' date, even if it looks and smells fine. Using it after this date could be a risk to health.

6	Storage conditions and/or conditions of use	Instructions on food labels should be carefully followed to ensure that food is safe and can be enjoyed at its best.
7	Name or business name and address of the food business operator	The name and address of the food business operator in the EU must appear on the label. The information provided should be detailed enough to enable a consumer to contact the business.
8	Place of origin or provenance (if implied)	Place of origin or provenance describes where food products come from and must be stated on a food label if the name suggests that it is from or has been made in a different country from where it was produced.
9	Food allergens	There are 14 food allergens, plus their derivatives, which by law must be highlighted within the ingredients list on a food product if they have been added deliberately. This is to enable consumers with a food allergy or food intolerance to make safe food choices.
10	Nutrition information	Nutrition information must appear on food labels in one of two formats: either energy value alone or energy value alongside fat, saturates, sugar and salt.
11	Alcohol strength	If a drink contains more than 1.2 per cent alcohol, the strength must be stated on the label.

Voluntary information on food labels and packaging

- Some information found on food labels is not required by law but is added by the food business **voluntarily** for marketing purposes or to provide consumers with useful information, such as serving suggestions or suitability for special diets (e.g. vegetarian/vegan). This information should not be misleading.

> **Voluntary**: not required by law.

Nutrition and health claims

- Nutrition claims on food labels are any claims that state, suggest or imply that the food has a particular beneficial nutritional property.
- A health claim is any claim that states, suggests or implies that a relationship exists between a food category, a food or one of its constituents and health.
- When food is fortified with micronutrients such as vitamins and minerals, this must be stated in the ingredients list.

Front-of-pack labelling

- A growing number of food manufacturers and supermarkets are using front-of-pack labelling to show at a glance what is in their food. While front-of-pack labelling is voluntary, when it is provided it must meet specific UK government guidelines regarding content and formatting.
- This is colour-coded red, amber and green:
  - Red means that the food is high in one of the nutrients that we should be cutting down on. These foods should be seen as treats and eaten in smaller amounts. A diet with fewer reds is healthier.
  - Amber means that the food is neither high nor low in the nutrient and is a suitable choice most of the time.
  - Green means that the food is low in the nutrient. The more green, the healthier the choice.
- Front-of-pack labelling also highlights percentage reference intakes – specific amounts of energy and nutrients that can be consumed on a daily basis in order to maintain a healthy diet. They show how much fat, saturated fat, salt, sugars and energy is in food per 100 g/100 ml.
- When front-of-pack labelling is used it must include the following:
  - energy value in kJ or kcals per 100 g/100 ml and in a specified portion of the product
  - information on the amount of fat, saturated fat, sugars and salt (in grams) in a specified portion of the product
  - portion size information that is meaningful for a consumer, for example one burger
  - percentage reference intake information based on the amount of energy and nutrients in a portion.

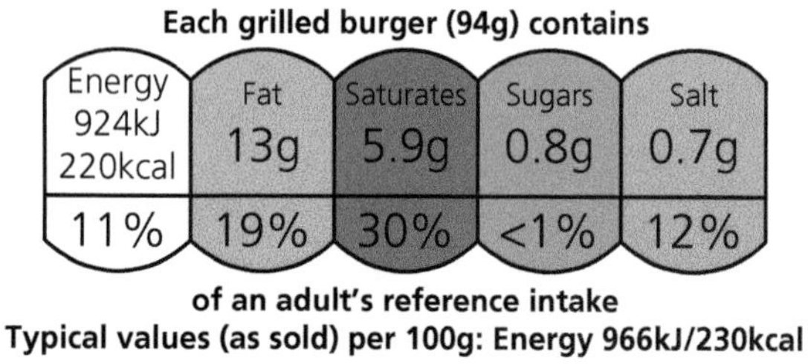

Each grilled burger (94g) contains

Energy 924kJ 220kcal	Fat 13g	Saturates 5.9g	Sugars 0.8g	Salt 0.7g
11%	19%	30%	<1%	12%

of an adult's reference intake
Typical values (as sold) per 100g: Energy 966kJ/230kcal

Figure 12.5 Colour-coded front-of-pack labelling

Marketing terms

- Consumers should be aware that manufacturers, producers and retailers use marketing terms to promote their food products. Examples of these terms include 'traditional', 'pure', 'homemade', 'natural' and 'fresh'.
- Such terms do not have agreed legal definitions and should not be used to mislead consumers. For example, a juice made from fruit concentrate cannot be described as 'fresh'.

Special dietary advice

- If a food is labelled vegetarian, it means that it does not contain any meat, including fish or poultry.
- Food products that display a 'suitable for vegetarians' logo or endorsement (e.g. from the Vegetarian Society) must meet specific requirements before such information can be used on a food label.
- To be labelled suitable for a vegan, food must not contain any ingredients of animal origin, including by-products such as gelatine.

Ethical and environmental food labelling schemes

- Consumers who consider ethical issues when shopping for food regard their purchasing power as a vote in favour of such issues.
- Ethical and environmental labelling, such as the Fairtrade Mark and the Soil Association organic symbol, can help consumers make an informed decision when shopping for food.

The Fairtrade Foundation

- The FAIRTRADE Mark is a registered certification label for food, drink, flowers, cotton and gold products sourced from producers in developing countries.
- When consumers purchase products with the FAIRTRADE Mark, they have the assurance that they are supporting producers to improve their quality of life and their communities. The FAIRTRADE Mark tells consumers that international Fairtrade Standards have been met with regards to the following:
  - Fair prices paid to producers – the 'Fairtrade Minimum Price' ensures that producers receive a consistent and reasonable price for the goods they produce. This price should cover all production costs and reflect the current market price for that product. The protection offered by Fairtrade will ensure farmers have a secure income so that they can plan for the future.
  - Community investment – the 'Fairtrade Premium' is paid into a collective fund to be invested in the community, enabling producers to improve their living and working conditions.
  - Safe and healthy working conditions – forced and child labour are prohibited and equality for all is encouraged. Fairtrade aims to ensure all workers are protected from abuse within the workplace.
  - Environmental protection – Fairtrade promotes environmental protection by offering training in areas such as resource management, recycling and effective waste management. Certain chemicals must not be used. Fairtrade also encourages an awareness of green farming techniques.

Figure 12.6 **The FAIRTRADE Mark**

The Soil Association

- The Soil Association sets standards for organic farming, food production, processing and sales.
- The standards are delivered through the Soil Association's wholly owned subsidiary, Soil Association Certification, which works with a wide range of food businesses, including suppliers, manufacturers, wholesalers and retailers.
- Food businesses that meet the standards and are certified by Soil Association Certification are allowed to use the Soil Association organic symbol on products they produce and sell.
- When you see the Soil Association organic symbol, you can be sure that what you buy has been produced to the highest standards. It means fewer pesticides, no artificial additives or preservatives, the highest standards of animal welfare and no GM ingredients.
- Consumers choosing products with this label can be assured that the food they are buying has been independently inspected and verified as organic and can be traced back to farm.

Figure 12.7 **The Soil Association organic symbol**

Food quality assurance schemes

- Food **quality assurance** schemes inform consumers that the food they buy has been produced safely and meets the highest standards of production.
- The Food Standards Agency monitors information and claims made by assurance schemes to ensure they are accurate.
- Members of a particular food quality assurance scheme can use the scheme's logo on their produce to advertise to consumers that the product has been produced to these standards.

The Northern Ireland Beef & Lamb Farm Quality Assurance Scheme

- The Northern Ireland Beef & Lamb Farm Quality Assurance Scheme (NIBL FQAS) is owned by the Livestock & Meat Commission (LMC) for Northern Ireland on behalf of the beef and sheep meat industry.
- The scheme was developed to give consumers assurances about the farm end of the production chain of their food.
- It is about farm quality – the quality of the production methods used, the quality of care for animals that is practised, the quality of the farm environment and, above all, the quality of concern for the customer in producing beef and lamb that is wholesome, safe and free from unnatural substances.
- LMC appoints a Certification Body to independently verify that producers are adhering to the required Standards of the scheme.
- NIBL FQAS is one of the longest established of all the farm quality assurance schemes, not only in the UK but probably worldwide, and is one of the strongest tools that the beef and sheep meat industries can have for marketing their products.

Information courtesy of the Livestock & Meat Commission for Northern Ireland

The Bord Bia Quality Mark

- The Bord Bia Quality Mark is a quality assurance scheme that:
  - informs consumers about the safety and traceability of food produced and processed in the Republic of Ireland
  - can be found on a wide range of food products, including meat, fruit, vegetables and eggs
  - covers food safety and standards from farm to fork with regards to animal health/welfare, traceability, food processing/manufacturing, farm safety/hygiene and management of the environment.
- Producers using the Bord Bia Quality Mark are regularly inspected to ensure they comply with the standards required by the scheme.

> **Quality assurance:** a system that guarantees food has been produced to meet specific standards of production.

Figure 12.8 **The Northern Ireland Beef & Lamb Farm Quality Assurance Scheme (NIBL FQAS) gives consumers assurances about the farm end of the food production chain**

Figure 12.9 **The Bord Bia Quality Mark**

> **Exam tip**
>
> Make sure you understand the difference between mandatory and voluntary food labelling, and can explain examples of each correctly.

> **Typical mistake**
>
> Candidates often confuse the factors affecting food choice – for example, mixing up personal and social factors. There can also be issues differentiating between store layout, financial incentives and advertising in examination responses.

Now test yourself

1 Which **one** of the following statements best describes the meaning of the Bord Bia Quality Mark (Figure 12.9)? [1 mark]
   A Suitable for vegetarians
   B Informs consumers about the traceability of food produced in the Republic of Ireland
   C Health claim
   D Value for money

2 Explain the impact of health issues on food choice. [3 marks]

3 Using the information on the label below, identify **two** pieces of mandatory information and **two** pieces of voluntary information. [4 marks]

4 Discuss **three** ways the layout of a supermarket can encourage a consumer to buy more products. [6 marks]

5 Assess a range of financial incentives used by retailers to influence food choice. [9 marks]

13 Food safety

There are many different types of **bacteria** and not all of them are dangerous. However, food must be prepared safely in order to reduce the risk of **pathogenic bacteria** contaminating food and causing **foodborne illness**.

Conditions needed for bacterial growth

REVISED

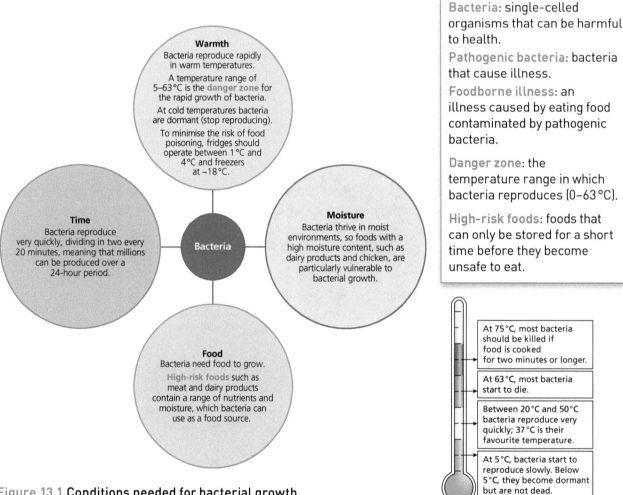

Figure 13.1 **Conditions needed for bacterial growth**

Bacteria: single-celled organisms that can be harmful to health.

Pathogenic bacteria: bacteria that cause illness.

Foodborne illness: an illness caused by eating food contaminated by pathogenic bacteria.

Danger zone: the temperature range in which bacteria reproduces (0–63 °C).

High-risk foods: foods that can only be stored for a short time before they become unsafe to eat.

Figure 13.2 **Different temperatures and bacteria growth**

Food poisoning bacteria

REVISED

- The symptoms, sources and methods of control for the most common food poisoning bacteria are given in Table 13.1.
- Pregnant women, babies and older adults are particularly vulnerable to food poisoning, because of issues they may have with their immune system.
- Extra care must be taken when preparing, cooking and serving food for these groups, to minimise their risk of foodborne illness.

Cross-contamination: when bacteria are transferred from one product to another, in particular from raw to cooked foods.

Table 13.1 **Symptoms, sources and methods of control for food poisoning bacteria**

Food poisoning bacteria	Symptoms	Food sources	Method of control
Campylobacter	• Diarrhoea (may be bloody) • Stomach cramps • Fever • Vomiting	• Raw or undercooked meat • Poultry • Unpasteurised milk • Untreated water	• Prevent cross-contamination • Cook meat and poultry thoroughly • Wash pre-packed salad leaves
E. coli	• Nausea • Vomiting • Diarrhoea (may be bloody) • Stomach cramps • Severe anaemia • Kidney failure (can be fatal)	• Raw meat • Gravy • Products made with unpasteurised milk (e.g. some types of cheese) • Raw fish	• Prevent cross-contamination • Cook meat and meat products (e.g. gravy) thoroughly • Choose pasteurised milk and cheese
Listeria	• Mild, flu-like symptoms • Nausea • Diarrhoea • Convulsions • Blood poisoning • Meningitis	• Unpasteurised cheeses • Soft, mould-ripened cheeses • Uncooked meats • Cold cuts of meat • Smoked salmon • Pâtés • Ready meals and ready-to-eat foods (e.g. pre-packed sandwiches)	• Cook meat and poultry thoroughly • Wash raw vegetables before eating • Keep uncooked meat separate from all other foods • Choose pasteurised milk and cheese • Prevent cross-contamination • Reheat ready-made foods until piping hot
Salmonella	• Fever • Diarrhoea • Vomiting • Abdominal pain	• Poultry • Raw meat • Eggs • Unpasteurised milk • Raw, unwashed vegetables	• Keep raw food, especially poultry, away from cooked and ready-to-eat foods • Ensure chicken, eggs and other meats are thoroughly cooked until piping hot in the centre to above 63 °C • Always wash hands after handling raw chicken and other raw meats
Staphylococcus aureus	• Severe vomiting • Diarrhoea • Abdominal pain	• Food made by hand that requires no cooking (e.g. prepared sandwiches, desserts and cream products) • Poultry • Cooked meats • Unpasteurised milk	• *Staphylococcus* bacteria often live on hands and in the nose, so high standards of personal hygiene are essential (e.g. wash hands after coughing or sneezing, and keep food handling to a minimum)

Food purchase

Table 13.2 **Dos and don'ts when buying food**

Do	Don't
✔ Observe the personal hygiene standards of staff handling and serving food ✔ Report unhygienic practices to store management or the local environmental health department ✔ Check that chilled and frozen food cabinets are operating at the correct temperatures and are not overloaded ✔ Buy chilled and frozen foods last and pack them together, preferably in an insulated bag or cool box ✔ Get chilled and frozen products home and stored appropriately as quickly as possible ✔ Pack raw foods, fruit and vegetables away from cooked and ready-to-eat foods	✘ Buy cans or packets of food that are damaged or have been opened ✘ Buy products from counters where raw and cooked foods are displayed and stored together ✘ Buy products that have exceeded the 'use by' or 'best before' date ✘ Delay in storing chilled and frozen food quickly and correctly

Food storage

- Storing food correctly is an important part of reducing the risk of food poisoning.
- Follow instructions to make sure that foods are stored in the correct place, at the correct temperature, for the correct length of time.
- Fresh foods that should be refrigerated include dairy products, meat, fish and ready meals, which should be stored in the fridge (1–4 °C).
- It is important to follow additional instructions regarding how long food can safely be kept in the fridge once opened.
- Leftovers should be stored in the fridge and used within two days.
- Storing food in a freezer (–18 °C) extends the shelf life of a wide range of foods, as very low temperatures prevent bacteria from multiplying.
- Bacteria will reproduce once food has defrosted, which is why thawed foods should not be refrozen.
- Food being stored in a freezer should be carefully wrapped and labelled.
- Food that does not need to be stored in a fridge or freezer to keep it safe to eat should be stored in a cool, dry storage area to maintain food quality.

Date marks

Most pre-packed food has either a use by or a best before date. Understanding these date marks is an important part of storing foods correctly.

Table 13.3 **Use by and best before dates**

Use by	Best before
• A use by date appears on the labels of highly perishable or high-risk foods. • These foods go off quickly and generally must be stored in a fridge or freezer that is operating at the correct temperature. • Food must be eaten by the use by date; after this date the food is likely to become unsafe to eat and could cause food poisoning.	• A best before date will generally appear on the label of low-risk foods that can be safely stored in a cupboard. • This date indicates how long the food will be at its best. • Most foods will remain safe to eat after this date, but the quality may be affected.

Personal hygiene

REVISED

The following rules apply when preparing, cooking and serving food:

- wash hands regularly, especially before preparing food and eating, and after handling raw foods, going to the toilet, touching waste, coughing and sneezing or handling pets
- avoid handling food if you are unwell or caring for someone who is unwell
- cover cuts, sores and burns with clean dressings or blue plasters and change these regularly
- wear a clean apron; never prepare food in unclean clothing
- remove jewellery before preparing food
- avoid touching hair and tie long hair back

Food preparation and cooking

REVISED

When preparing and cooking food, remember to:

- wash and dry hands thoroughly before handling food
- keep raw and cooked food separate at all times
- ideally use separate chopping boards for raw and cooked foods
- wash root vegetables such as potatoes, leeks and carrots thoroughly before use, as they often have traces of soil on them
- avoid preparing food for yourself or others if you are ill
- always follow instructions carefully when defrosting and cooking pre-packaged frozen foods
- always follow label instructions and recipes for cooking times and temperatures
- cook all foods until they are piping hot
- keep cooked foods covered and piping hot, above 63 °C, until it is time to eat them
- keep prepared cold foods in the fridge until it is time to eat them
- reheat cooked foods to a core temperature of 70 °C for at least two minutes
- do not reheat food more than once.

Environmental health practitioner

Every council has an environmental health department, which employs environmental health practitioners (EHPs) who try to make sure that the food we eat is safe.

Table 13.4 **Role of environmental health practitioners**

Role	Details
Inspecting food premises	EHPs inspect all premises where food is manufactured or handled to ensure appropriate food safety measures are in place.
Enforcing food legislation	EHPs enforce relevant food safety legislation including: • Food Safety (Northern Ireland) Order 1991 • Food Hygiene Regulations (Northern Ireland) 2006 • Food Information Regulations (Northern Ireland) 2014.
Establishing food standards	EHPs create food sampling plans and regularly remove samples of food on sale to the public so that they can be tested to detect the presence of toxins, contaminants, allergens and food poisoning bacteria.
Advising and educating food-based businesses	EHPs provide free advice and support to businesses within their council area.
Hygiene education	EHPs promote initiatives to prevent food poisoning at home. They also organise training for people working in the food industry, with successful candidates achieving a recognised food hygiene qualification.
Investigating food complaints	EHPs investigate complaints from consumers regarding foodborne illness, foreign objects in food, and food labelling.
Investigating food poisoning outbreaks	EHPs investigate cases of food borne illness reported by the Department of Health and members of the public.
Responding to food alerts from the Food Standards Agency	EHPs are involved in monitoring food incidents, hazards and alerts from the Food Standards Agency in order to protect consumers against foodborne illness and allergens.
Checking food labelling does not mislead the consumer	EHPs sample a wide range of foods to protect consumers against dishonest labelling and inaccurate descriptions of food, which can be classified as food fraud.

The Food Standards Agency's Food Hygiene Rating Scheme

- The Food Hygiene Rating Act (Northern Ireland) 2016 introduced a new statutory food hygiene rating scheme in Northern Ireland, making the implementation of the Food Hygiene Rating Scheme (FHRS) by district councils and the display of food hygiene ratings mandatory.
- The scheme helps consumers to make informed choices about where they eat or shop for food by highlighting how seriously a food business takes its food hygiene standards.
- The scheme is run by local councils and applies to restaurants, pubs, cafés, takeaways, hotels, supermarkets and other food shops.
- Food businesses are given a hygiene rating after inspection from a food safety officer from their local council, who checks for compliance with food hygiene legislation. At the end of the inspection, the business is given one of six different ratings from 0 to 5.
- Areas covered by the inspection may include:
  - how hygienically the food is handled – how it is prepared, cooked, reheated, cooled and stored
  - the condition of the structure of the buildings – the cleanliness, layout, lighting, ventilation and other facilities
  - how the business manages and records what it does to make sure food is safe.

Table 13.5 **Food Hygiene Rating Scheme ratings**

Rating	
0	Urgent improvement necessary
1	Major improvement necessary
2	Improvement necessary
3	Generally satisfactory
4	Good
5	Very good

Figure 13.3 **Food hygiene rating sticker**

Food Safety (Northern Ireland) Order 1991

- The Food Safety (Northern Ireland) Order 1991 protects consumers from poor standards of food hygiene and the risk of food borne illness. The law is enforced by environmental health practitioners.
- The Order makes it an offence to produce, treat or alter food in any way that could endanger health or to sell food that:
  - ○ is unfit for human consumption
  - ○ has been declared injurious to health
  - ○ is so contaminated that it would be unreasonable to expect consumers to eat it in that state
  - ○ is not of the nature, substance or quality demanded by the consumer
  - ○ is falsely or misleadingly labelled or presented.
- The Order provides enforcement authorities with powers to:
  - ○ issue improvement notices
  - ○ inspect and seize suspect food
  - ○ issue emergency prohibition notices and orders.

The Food Hygiene Regulations (Northern Ireland) 2006

- This food safety legislation assumes that all food sold to consumers has been produced, processed or distributed in accordance with specific hygiene regulations and is fit for human consumption. It is enforced by the Food Standards Agency and environmental health practitioners.
- Examples of specific regulations within this legislation include:
  - ○ cold holding requirements that make it an offence to store food that is likely to support the growth of pathogenic micro-organisms at a temperature above 8 °C
  - ○ hot holding requirements that make it an offence to store food that is likely to support the growth of pathogenic micro-organisms at a temperature below 63 °C
  - ○ restrictions on the sale of raw milk (selling raw milk for direct human consumption is in breach of the regulations)
  - ○ requirements for local retailers supplying small quantities of meat direct from farms (e.g. a label indicating the name and address of the farm where it was reared and slaughtered)
  - ○ the requirement that food businesses have Hazard Analysis and Critical Control Points (HACCP) procedures in place, which ensures that they have taken appropriate steps to identify and control potential food safety risks within the food production process.

Exam tip

Make sure you know the names of the food poisoning bacteria and how to spell them correctly. Learn the typical food sources of each – for example, *E. coli* in raw meat and fish.

Typical mistake

Create a visual aid to help you remember the different food safety temperatures as many candidates lose easy marks by providing incorrect figures.

Now test yourself

TESTED

1 Who is responsible for enforcing The Food Hygiene Regulations (Northern Ireland) 2006?   [1 mark]

2 Write down **two** food safety points for storing raw meat to reduce the risk of food poisoning.   [2 marks]

3 Explain why there is a 'use by' date on pre-packed sliced cooked meat.   [3 marks]

4 Discuss how an environmental health practitioner protects the consumer in keeping food safe.   [6 marks]

5 Explain the personal hygiene advice that should be followed when preparing, cooking and serving food.   [9 marks]

14 Resource management

Food is one of the most important resources on the planet, and we have a responsibility to use it efficiently and economically to reduce food waste, save money and protect the environment.

Managing time, energy and money for food choice, food shopping, food preparation and food storage

REVISED ☐

Resource management refers to how we manage our time, energy and money. This can be affected by a number of factors, including income, personal circumstances and family size.

Table 14.1 **Factors affecting resource management**

Food choice	Food shopping	Food preparation	Food storage
Choose own-brand food items that can often offer better value for money	Prepare a shopping list to avoid impulse buys	Use leftovers (e.g. leftover roasted vegetables could be used in a frittata or wrap)	Store leftovers appropriately (e.g. in the fridge) and use within two days, or freeze and use within one month to eliminate food waste
Choose local and seasonal food to reduce the amount of energy used during production, transportation and storage	Price check between food stores to ensure value for money	Prepare and cook meals from scratch to avoid the price premium of convenience foods and ready meals	Consider the type of ingredient (e.g. frozen or tinned vegetables that can be stored until needed)
Consider what form of food offers the best value for money (e.g. choosing a whole chicken is often cheaper than chicken portions)	Consider shopping options that might save time (e.g. shopping lists can be saved for future use when shopping online)	Choose the most energy-efficient cooking method (e.g. bake potatoes using a microwave rather than an oven)	Store all food correctly in the fridge, freezer or cupboard to extend shelf life (e.g. bananas can cause other fruit to ripen rapidly, so store separately)
Choose alternative ingredients that can be cheaper and quicker to cook (e.g. make a lentil or Quorn Bolognese instead of a meat-based sauce)	Compare unit prices to ensure you are getting the best deal (e.g. bulk buys may not always offer the best value for money)	Choose the most energy-time-efficient appliances (e.g. slow cooker or food processor)	Check date marks regularly for foods approaching their use by or best before date and try to use them up
Plan meals carefully and shop once a week or fortnight to save time and effort	Avoid special offers (e.g. buy one get one free) unless they provide real value for money and the food can be used	Cook and serve the correct portions for your specific needs	Keep food wrapped or stored in appropriate packaging to extend shelf life (e.g. store fresh herbs in a sealed plastic bag)

Strategies to reduce food waste

- When we throw food away, we are wasting resources as well as the food itself – for example, the water, fuel and labour used to grow, harvest, store and transport the food.
- Consumers have a responsibility to manage money wisely and avoid unnecessary or wasteful buying. They could consider some of the following strategies:
  - plan meals and write a shopping list to ensure you only buy what you need
  - be accurate with portion sizes, e.g. a portion of pasta for an adult = 100 g
  - buy food with the longest use by date so you have time to use it
  - make the most of food purchased by using leftovers for a second meal
  - buy ingredients that are versatile and can be used in a range of different meals
  - keep a range of useful store cupboard essentials that can be used to create meals quickly, such as dried herbs/spices, tinned tomatoes, pasta, rice, flour, couscous, noodles, stock cubes, beans, tuna and oils/vinegars
  - store food correctly to keep it fresher for longer, e.g. place fruit and vegetables in the salad drawer of the fridge, and store potatoes and onions in a dark cupboard
  - freeze food before the use by date, defrost when needed and use within 24 hours
  - cook and eat first the food that was bought first; store newly bought food at the back of fridge, freezer or cupboard to make this easier to achieve
  - avoid **marketing strategies** when shopping for food, which can often lead to impulse buys and generate food waste.

> **Resource management:** how we use resources such as time, energy and money.
>
> **Seasonal food:** food that is ready to harvest and at its best at a particular time of year.
>
> **Marketing strategies:** promotional activities used by retailers to increase food sales.

Figure 14.1 **Storing food correctly can help to reduce food waste**

Table 14.2 **Methods of paying for food**

Method	Advantages	Disadvantages
Cash	• Paying for food with cash is quick, efficient and cost effective. • It is easier to track spending on food and stay within a budget, as you can only spend the amount of cash you have available.	• Disadvantages of paying for food with cash often relate to accessing cash from your bank account. • There are security issues associated with carrying large amounts of cash.
Credit card (including contactless) A credit card enables a consumer to borrow money from the card issuer to pay for products or services. Card holders receive regular statements of how much they have borrowed and need to repay.	• Buying food with a credit card can enable consumers to receive benefits linked to that card, such as cashback or rewards/loyalty points. • Using a credit card means you do not need to carry large sums of cash and, if the bill is paid in full at the end of each month, it offers interest-free borrowing. • Contactless credit cards offer convenience at the checkout as they speed up the payment process.	• Credit cards are a form of borrowing; it can be a high-risk decision to use a credit card to meet basic needs such as food. • If the credit card bill is not paid in full at the end of the month interest is applied, increasing the cost of the food bill and the risk of debt. • Concerns exist regarding the security of contactless payment methods, as consumers do not usually have to enter security information (e.g. PIN).
Debit card (including contactless) A debit card draws from the funds in your bank account to pay for products purchased or services used.	• Debit cards are a quick and efficient method of paying for food. • They are cost effective as they do not usually incur charges or interest for the consumer. • Some debit cards allow consumers to request cashback. • Contactless debit cards offer convenience at the checkout.	• As debit cards only allow consumers to spend the amount of money they have available in their account, consumers must ensure their accounts are in credit. • Concerns exist regarding the security of contactless payment methods, as consumers do not usually have to enter security information (e.g. PIN) • As there is a maximum amount of £30 allowed per contactless transaction, it may not always be possible to use this method when shopping for food.

Contactless: a function on some debit/credit cards that allows quick and easy payments of up to £30 without entering a PIN.

Interest: the charges associated with borrowing money, including fees.

Debt: a sum of money that is owed.

PIN: Personal Identification Number; used as a security measure when using credit or debit cards.

Figure 14.2 **Many food retailers now accept contactless credit/debit cards**

Exam tip

When answering questions that ask you to 'explain', remember to write in detail and give appropriate examples. For example, if you are asked to explain strategies for reducing food waste, include relevant examples in your answer.

Typical mistake

Answers can be general in nature with examples that are not relevant to the specific question – for example, is the question about food choice, food shopping, food preparation or food storage?

Now test yourself

TESTED

1 Identify **two** reasons for food waste. [2 marks]
2 Explain **two** strategies someone on a low income could use to manage their spending when shopping for food. [4 marks]
3 Explain **three** strategies that families could use to reduce food waste. [6 marks]
4 Evaluate the use of a contactless debit card when paying for a takeaway coffee. [6 marks]
5 Evaluate the use of a credit card when shopping for food. [9 marks]

Success in the examination

In GCSE Home Economics: Food and Nutrition you will be assessed against three assessment objectives (AOs).

Assessment objectives		Food and Nutrition External Assessment (Component 1 – written examination)
AO1	Recall, select and communicate knowledge and understanding of a range of contexts.	15%
AO2	Apply skills, knowledge and understanding in a variety of contexts, and in planning and carrying out investigations and tasks.	20%
AO3	Analyse and evaluate information, sources and evidence, make reasoned judgements and present conclusions.	15%
Total weighting		50%

When will the exam be completed?

This is a linear qualification, with an external written exam at the end of the two-year course.

The written exam is worth 50 per cent of your GCSE Home Economics: Food and Nutrition qualification.

Assessment is available each summer from 2019.

How long will I have to complete the exam?

You will have **two hours** to complete the external written examination.

What type of questions will appear in the exam paper?

The written paper includes the following types of questions:

Multiple-choice questions

> **Example**
>
> Which **one** of the following foods is an example of a
> high biological value protein? [1 mark]
> A Eggs
> B Oranges
> C Tomatoes
> D Beans

Short response questions

> **Example**
>
> Give **one** example of a food that is caught. [1 mark]

Structured response questions

> **Example**
>
> Explain **two** reasons why it is important for teenagers
> to drink plenty of water. [4 marks]

Extended response questions

> **Example**
>
> Evaluate an adult's decision to do their food shopping
> online. [9 marks]

Tips on preparing for the exam

Examination success depends on a range of factors and it is important to prepare fully to achieve your best:

- Your notes should be carefully organised and review each of Chapters 1–14 of the course.
- Make succinct and focused revision notes for each chapter as you progress throughout the course – these can be used to prepare for class tests, school examinations and the final external examination (see tips for making effective notes below).
- Revision should be a regular part of your weekly routine over this two-year course of study.
- Effective revision should include practising examination-style questions.
- Know and understand the relevant command words that are used in the examination:

Command word	Meaning
Analyse	Separate information and identify characteristics
Assess	Make an informed judgement
Calculate	Work out the value of something
Define	State a precise meaning
Describe	Set out characteristics
Discuss	Present key points
Evaluate	Consider evidence for and against
Examine	Investigate closely
Explain	Set out purposes or reasons
Explore	Investigate without preconceptions
Identify	Name
Justify	Support with evidence
Outline	Set out main characteristics
Suggest	Present ideas for consideration

Tips for making effective notes

- Be selective and write in your own words.
- Keep notes well spaced so that you can see individual points and add in more details later if necessary.
- Make links between topics.
- Add illustrations, examples and diagrams to put ideas into a practical context.
- Make notes memorable using colour, highlighting and underlining.
- Read through your notes to make sure that they are clear and accurate.
- Use command words to determine depth of notes.
- Be active in your approach – copying is not an effective way to make notes.
- Complete notes early so that they can inform other revision strategies.

Approaching the paper

Consider the following tips to help maximise your examination mark:

- The exam paper has a total of 120 marks. As you have two hours to complete the examination paper, this means you need to spend **one minute on every mark**.
- Read the instructions on the front of the examination paper carefully.
- The key to scoring high marks in an exam is always by reading the question carefully.
- The paper will have questions from all aspects of the specification, so be prepared.
- Make sure you understand the meaning of all command words and demonstrate this skill to address the focus of the question.
- Consider the mark allocation for each question.
- In short response questions, aim to complete one sentence for each mark.
- Use subject-specific key vocabulary to access higher marks.
- Keep your answer focused to the question – try underlining key words in the question.
- Avoid repetition as you will not be awarded marks twice.
- Quality of written communication (QWC) will be assessed in extended response questions and this will be highlighted on the front cover of the examination.

Quality of written communication

In GCSE Home Economics: Food and Nutrition, you must demonstrate your quality of written communication. QWC is assessed in responses to questions and tasks that require extended writing. You need to:

- ensure that text is legible, and that spelling, punctuation and grammar are accurate, so that meaning is clear
- select and use a form and style of writing that suits the purpose and complex subject matter
- organise information clearly and coherently, using specialist vocabulary where appropriate.

Sample examination questions

In this section you will see some examples of the types of questions you may get in your written examination. They show some candidates' responses and there is an explanation of why the marks were awarded.

Example

Outline the effect of having too much vitamin A in the diet. [2 marks]

Candidate response 1

Excess vitamin A is dangerous as it can be stored in the liver. It is particularly dangerous for pregnant women as it can cause birth defects.

Candidate response 2

Having too much vitamin A is bad as it can affect the teeth and makes pregnant women unwell.

Assessment comment

- To achieve top marks candidates should aim to present one point for each mark.
- In the first response the candidate has presented two factually accurate points, making good use of key terms. This response achieved two marks.
- In the second response the candidate has presented inaccurate and vague information, making limited use of key terms. This response achieved no marks.

Example

Explain how eating more fruit and vegetables benefits health. [3 marks]

Candidate response 1

Fruit and vegetables are important for our health as they are good for us. They have lots of different nutrients, which can keep us healthy. They contain vitamin C, which helps to prevent colds and flu.

Candidate response 2

Eating fruit and vegetables can benefit our health as they are high in dietary fibre, which can help keep the digestive system healthy and reduce the risk of bowel disorders such as constipation. They are also low in calories and should be eaten regularly to help with weight management and to reduce the risk of obesity.

Assessment comment

- To achieve top marks candidates should aim to present one point for each mark.
- In the first response the candidate has presented two points of general knowledge that lack factual detail. In the final sentence they have provided basic subject-specific knowledge. This response achieved one mark.
- In the second response the candidate has used the wording of the question to focus their answer. They have provided three different points and used key terms accurately. This response achieved three marks.

Example

Explain **two** food safety tips that should be followed when preparing and cooking chicken.　　　[4 marks]

Candidate response 1

1　To avoid cross contamination, you should use separate chopping boards and knives when preparing raw chicken.

2　To ensure chicken is cooked, check that it is piping hot with no pink meat and juices running clear.

Candidate response 2

1　Use a red board when preparing chicken.

2　Wear a clean apron.

Assessment comment

- To achieve top marks candidates should aim to present one point for each mark.
- In the first response the candidate presented two well-explained points that relate directly to the preparation and cooking of chicken and include relevant key terms. This response achieved four marks.
- In the second response the candidate has not explained the points. The second answer relates to personal hygiene and not the preparation and cooking of chicken. This response achieved one mark.

Example

Discuss **three** dietary and lifestyle factors that may lead to the development of osteoporosis.　　　[6 marks]

Candidate response 1

Not having enough calcium in your diet can lead to osteoporosis and this makes your bones break more easily. You should also have some Vitamin C in your diet to help your body absorb more calcium from all the food you are eating. You should eat foods with calcium and Vitamin C every day.

Candidate response 2

It is essential to have a good supply of calcium-rich foods such as dairy products in your diet as this will strengthen the bones and reduce the risk of osteoporosis in later life. With this you should also consume a range of foods rich in Vitamin D, for example fortified cereals, as this encourages the absorption of calcium from foods. Finally, you should take regular weight-bearing exercise, for example, jogging or brisk walking, as this is good for strengthening bones.

Assessment comment

- In the first response the candidate presented incorrect information – vitamin C does not help the body absorb more calcium. Discussion only included information on dietary factors with no mention of lifestyle and was repetitive. This response achieved two marks.
- In the second response the candidate clearly states both dietary and lifestyle factors. Information is accurate and discussed in relation to osteoporosis, with relevant examples. This response achieved six marks.

Example

Using the Eatwell Guide, suggest and justify a breakfast to meet the nutritional needs of an active adolescent (12–18 years). [12 marks]

Candidate response 1

Breakfast choice: Scrambled egg, wholemeal toast and pure orange juice.

The breakfast I have chosen will provide a range of nutrients that are particularly important for adolescents. The scrambled egg contains high biological value protein, which this is essential for an adolescent for growth and for repair of tissues. It can also help with maintenance of muscle, which is important for teenage boys. Scrambled egg is made using some milk and this is a good way of including some calcium. Teenagers need a good supply of calcium for their bones and to achieve peak bone mass.

Breakfast should contribute 25 per cent of a teenager's energy for the day. Energy from the wholemeal bread will help meet the energy requirements of an active teenager. They are also going through a period of rapid growth and need energy for this too. The fact that wholemeal toast is higher in fibre compared to white bread means that they will be less likely to suffer from constipation and it will keep their digestive system healthy. Wholemeal toast is filling and this will help ensure that the teenager does not feel the need to snack before break time. If there is margarine on the toast this will provide some fat for the teenager. Vitamin A found in the margarine is important at this stage for healthy skin.

There will be iron in the egg yolk. This is required for the production of haemoglobin to transport oxygen to muscles to boost energy and avoid muscle fatigue, which is especially important for teenagers who are active. The pure orange juice will provide one of the teenager's five a day as well as vitamin C, which is needed to absorb iron.

The pure orange juice is also a source of fluid and this can prevent dehydration as a result of sweat lost during physical activity. Teenagers who skip breakfast are more likely to be dehydrated and this will make it difficult to concentrate and perform well at school.

Candidate response 2

The teenager should make sure they have a bowl of cereal for breakfast before they leave the house. Cereal is good for you and keeps you healthy. It will help to get you going in the morning and gives you energy. Teenagers need to be careful not to skip breakfast as breakfast is the most important meal of the day. Lots of teenagers skip breakfast and this is not healthy. To be healthy you should have breakfast. The cereal will give the teenager energy and the milk will give them calcium. This is good for strong bones and it is important to have healthy bones for the rest of your life. Sugary stuff will ruin your teeth and the teenager should have a cereal that isn't sugary. A teenager should have cornflakes instead of Frosties.

Assessment comment

- In the first response the candidate presented a suitable breakfast for an active adolescent and included a range of food items from each section of the Eatwell Guide. Ingredients were appropriately combined and a drink was included. The overall impression of the response is highly competent. The candidate shows excellent knowledge and understanding of the Eatwell Guide as well as the nutritional needs of an active adolescent. They have identified and commented on an excellent range of key points relevant to the question, and the quality of written communication is highly competent. This response achieved Level 3: 9–12 marks.
- In the second response the candidate presented a basic breakfast. However, they did not include a range of food items from each section of the Eatwell Guide. Ingredients were vague and a drink was not included. The overall impression of the response is basic. The response shows limited knowledge and understanding of the Eatwell Guide and nutritional needs. They have identified and commented on a few obvious points relevant to the question and the quality of written communication is basic. This response achieved Level 1: 1–4 marks.

Now test yourself answers

1 Food provenance

1 Choose **two** from: poultry (e.g. chicken, turkey), beef and veal, pork, mutton and lamb, goat and kid, and game (e.g. rabbit, venison).

2 Consumers may choose to buy sustainable fish to prevent overfishing and the decline of wild stocks.

3 Choose **three** from: preparing soil, sowing seeds/seedlings, watering, fertilising, weeding, protecting from pests, harvesting, separation and inspection, and storage.

4 (a) Organic farming is characterised by farming methods that protect the environment, e.g. hand weeding and the use of green manure. Animal welfare is considered and animals are given space to move freely.

   (b) Intensive farming is characterised by farming methods that often rely on machinery to produce high yields. For example, pesticides are used to control pests and chemical fertilisers are used to enrich soil. Animals are kept indoors with limited space to move around, e.g. caged chickens.

5 It is important that consumers consider food provenance when shopping for food so that they can identify where food comes from. This allows them to choose specific food products, for example, locally-grown vegetables. It also means that they can choose food in season, which often has a better flavour, for example, strawberries in summer. Food provenance also increases food traceability, as food can be traced from farm to fork, for example, NI meat.

2 Food processing and production

1 Choose **three** from: weighing and measuring, mixing, proving, shaping, baking, slicing.

2 Cows are milked at least twice a day. Milk is stored at 4 °C and transported for processing at a dairy. It is then pasteurised to destroy pathogenic bacteria, before being separated into the cream and liquid components. These components are then reblended depending on the type of milk being produced, e.g. whole, semi-skimmed or skimmed milk. Finally, milk is homogenised to achieve an even consistency.

3 The food supply chain starts in the agricultural sector, where food is grown, reared or caught.

Food then goes through primary and secondary processing within the manufacturing sector. Food is then supplied to businesses using a range of different transport methods within the distribution sector. Finally, food and drink products are available to consumers from a range of outlets.

4 Answer could include:
- antioxidants are used to extend shelf life, e.g. bakery products
- colours make food look more attractive, e.g. confectionery
- emulsifiers prevent ingredients from separating during storage, e.g. mayonnaise
- flavourings replace flavour lost during processing, e.g. chilled meals
- preservatives keep food safe to eat for longer, e.g. cured meats
- sweeteners make food taste sweet, e.g. low-calorie drinks.

5 Answer could include:
- to enrich products for individuals with special diets, e.g. soya milk fortified with B12 to meet the nutritional needs of vegans
- to enrich a staple food with a nutrient that it does not naturally contain, e.g. vitamin A is added to margarine so it matches the nutritional value of butter
- to improve the nutritional status of a specific group of people who may be deficient in a particular nutrient, e.g. fortified breakfast cereals can help children meet their nutritional needs.

6 Advantages of additives include:
- they keep food safer for longer
- they provide a wider variety of foods to choose from
- they ensure food is enjoyable to eat.

Disadvantages of additives include:
- they can make low-quality products seem better than they are
- they may affect some children's behaviour
- they may have a negative impact on health.

3 Food and nutrition for good health

1 6 g

2 Any suitable choice, for example, **two** from: margarine, olive oil, rapeseed oil.

3 Main nutrients supplied by the five food groups from the Eatwell Plate:
- Potatoes, bread, rice, pasta and other starchy carbohydrates = carbohydrate
- Fruit and vegetables = vitamin C
- Beans, pulses, fish, eggs meat and other proteins = protein
- Dairy and alternatives = calcium
- Oil and spreads = fat

4 Beans and pulses are high in fibre and can have a positive impact on bowel health. They are also filling and low in calories, which can help us maintain a healthy weight.

5 We should be eating less red and processed meat because it is high in saturated fat and salt. Saturated fat increases blood cholesterol levels and the risk of heart disease. High salt intake can increase blood pressure, which leads to hypertension, a risk factor for heart disease and stroke.

6 (a) It is important to follow the tip 'Don't get thirsty' to prevent dehydration and maintain body temperature. Water is also important in the diet to prevent constipation and maintain blood pressure.

(b) It is important to follow the tip 'Eat lots of fruit and vegetables' as they are a good source of soluble fibre, which helps to prevent constipation. They are also a good source of antioxidant nutrients (vitamin C), which reduces the risk of heart disease and cancer.

7 Discussion of the Eatwell Guide as a tool for achieving a balanced diet may include the following points:
- Use advice from the Eatwell Guide when shopping for food, cooking at home, eating out and planning meals, to increase health and well-being.
- It is a useful tool for achieving a balanced diet as it shows the different types of foods and drinks we should consume by providing colourful images of a wide range of foods for each food group.
- It applies to everyone, except children under two years of age, as they have different nutritional requirements.
- It highlights what proportions of each food group we should consume for good health, for example, two-thirds of our diet should be made up of fruit and vegetables, and starchy carbohydrates.
- There is also specific advice regarding each food group, for example, eat at least five portions of a variety of fruit and vegetables every day as they are a good source of soluble fibre, which helps to prevent constipation.
- Oils and spreads make up the smallest group and should be eaten less often and in smaller

amounts. Excess consumption of fat has a negative impact on health because it increases the risks of obesity and heart disease.
- The Eatwell Guide also provides advice about hydration – we should drink six to eight glasses of fluid per day to prevent dehydration and constipation.
- The Eatwell Guide advises that we should eat foods high in sugar and saturated fat less often and in small amounts to reduce the risk of diet-related disorders, for example type 2 diabetes and tooth decay.
- The Eatwell Guide provides a reminder to check the label on packaged food in order to choose foods that are lower in fat, salt and sugars and to achieve a balanced diet.

4 Energy and nutrients

1 1 gram of carbohydrate provides 3.75 kilocalories.

2 Any **two** from: fruit, vegetables, low-fat dairy, fish, wholegrain cereals.

3 Answer to include **two** from:
- older adults can achieve energy balance and maintain a healthy weight by consuming foods that are nutrient dense rather than energy dense, for example they should snack on fruit rather than biscuits or sweets
- ageing reduces energy needs because growth has stopped and physical activity levels may have declined, so it is essential to balance energy intake from food to maintain a healthy weight
- older adults should try to maintain physical activity levels (PAL) as a way to achieve energy balance by expending kilocalories, for example by walking, swimming or gardening.

4 (a) Males aged 19–24.

(b) If people continue to eat the same number of calories at 65 as they did at 25 they are likely to gain weight, as BMR declines with advancing age. Levels of physical activity may also be reduced, which reduces the amount of kilocalories expended and contributes to weight gain.

(c) Male and female energy requirements vary as men tend to have greater muscle mass, which requires more energy. Men who sustain levels of physical activity or have a manual job are more likely to need more energy than women, who may be more sedentary in terms of regular participation in sport and type of job.

5 The following factors have an impact on energy requirements:
- Basal metabolic rate (BMR) accounts for 75 per cent of a person's energy needs. It is the amount of energy your body needs to maintain functions

CCEA GCSE Home Economics: Food and Nutrition 89

such as breathing and to keep at a constant temperature when totally at rest, e.g. sleeping.
- Specific need – female energy requirements increase slightly to meet the demands of pregnancy and breast feeding (lactation).
- Thermogenic effect of food – this refers to an increase in energy expenditure after eating, while the body is digesting food.

5 Macronutrients

1 No more than 35 per cent of dietary energy should come from total fat.

2 Answer to include **two** from the following: pulses, beans, grains, nuts, soya, tempeh, Quorn (if specified for vegan).

3 Answer to include **two** from: all sugars added to food by the manufacturer, cook or consumer, and sugars present in honey, syrups and unsweetened fruit.

4 When two LBV protein foods are eaten together, e.g. beans on toast, the amino acids in one food will compensate for the limited amino acids of the other, resulting in a meal of high biological value. This is known as complementation (the complementary action of proteins).

5 It is important to include unsaturated fat in the diet because omega 3 fatty acids can help prevent blood clotting, which protects the heart and can reduce the risk of cardiovascular disease. Omega 6 has a positive impact on blood cholesterol and can also reduce the risk of cardiovascular disease.

6 Starchy carbohydrates are important in the diet of an adolescent as they are a good source of energy, which is important at this stage of rapid growth. A lack of carbohydrates in the diet means protein will be used as a source of energy instead of being used to maintain and repair tissue. Starchy carbohydrates are also important to provide energy to meet the increased demands of physical activity common during adolescence.

6 Micronutrients

1 Three foods that are a valuable source of vitamin C:

Oranges	X	Cheese	
Kiwi fruit	X	Red peppers	X
Eggs		White bread	

2 Answer to include **two** from: table salt, salty snacks (e.g. crisps), processed foods (e.g. processed meat products), breakfast cereals, cheese.

3 A deficiency of vitamin D leads to skeletal deformity called rickets in children and osteomalacia in adults.

4 Answer to include **two** from: non-haem iron from plant sources is not as easily absorbed as haem iron from animal sources. Absorption of iron is reduced by phytates in cereals and tannins in tea, which make it more difficult to absorb iron from food.

5 Answer could include any **two** from:
- chicken breast could be coated in breadcrumbs as white bread is fortified with calcium
- a white sauce could be added, as milk and flour are sources of calcium
- peas could be replaced with dark green leafy vegetables, e.g. cabbage, as these are a source of calcium.

7 Fibre

1 Answer to include **one** from:
- bowel disorders, e.g. constipation, bowel cancer
- heart disease/raised blood cholesterol/high blood pressure
- diabetes
- excessive weight gain/obesity.

2 Answer to include **two** from:
- grains (oats, barley, rye)
- pulses (peas, beans, lentils)
- fruit (apples, banana, pears)
- vegetables (carrots, parsnips, potatoes).

3 Answer to include **two** from:
- add vegetables, e.g. red pepper
- add pulses, e.g. kidney beans
- swap white rice for brown rice.

4 Answer to include **three** from:
- choose a wholegrain cereal for breakfast, e.g. Weetabix
- swap white bread for brown/multigrain/wholemeal bread
- eat more pulses, e.g. peas, beans, lentils
- swap white pasta for wholewheat pasta
- swap white rice for brown rice
- eat more fruit and vegetables, especially with skin on, e.g. pears.

5 Insoluble fibre assists digestion by enabling the body to get rid of waste more effectively. It helps prevent constipation by adding bulk to faeces, making them easier to pass. Insoluble fibre is filling and can reduce the desire to eat between meals, helping to maintain a healthy weight.

6 Soluble fibre benefits health as it makes stools soft and easier to pass, as soluble fibre dissolves in water and forms a gel in the digestive system. It can also help lower blood cholesterol levels, therefore reducing the risk of cardiovascular disease.

8 Water

1 Answers may include **three** from:
- water transports nutrients and oxygen through the body in blood
- water helps to maintain blood pressure
- water helps to prevent constipation, so contributes to bowel health
- water assists chemical reactions in the body, e.g. digestion
- water helps the kidneys filter waste, which is excreted as urine
- water provides fluid to keep joints mobile
- water is a key component of saliva, which helps swallowing
- water is a component of spinal fluid, which cushions the nervous system
- water helps to regulate body temperature via perspiration
- water helps to form tears to lubricate the eyes.

2 Six to eight glasses of water a day (between 1200 ml and 1600 ml).

3 Fluid is important in the diet of an older person because, as people age, their thirst sensation decreases. Older adults should sip at water throughout the day and have regular cups of tea/coffee to prevent dehydration.

4 In order to stay hydrated, it is important to develop a habit of regular fluid intake without waiting for thirst to intervene. Children should drink plenty of water because they do not always recognise the early stages of thirst and are less heat tolerant than adults. They are therefore more susceptible to dehydration in warm climates.

5 Any **two** from:
- drink more fluid when the weather is hot to stay hydrated
- increase water intake before, during and after physical activity to replace fluid lost by sweating
- include a drink at breakfast to rehydrate, e.g. water, fruit juice, tea/coffee, milk on cereal
- carry a water bottle with them in school and sip water regularly throughout the day.

9 Nutritional and dietary needs

1 Custard, jelly.

2 Calcium is essential in adolescence to help achieve peak bone mass and optimal bone health.

3 A food allergy to a specific food causes the body's immune system to react. Someone with a severe food allergy can experience a life-threatening reaction. Food intolerance does not involve the immune system and is generally not life-threatening.

4 (a) Protein intake should be maintained by older adults to help in the repair and maintenance of body tissues and to aid recovery from illness.

(b) Carbohydrate needs should be met by consuming starchy carbohydrates rather than sugary carbohydrates, which are high in calories but provide a limited range of other nutrients. Older adults should eat a range of fibre-rich foods to avoid constipation and other bowel-related disorders that are common at this life stage.

5 Answer could include: Pregnant women should follow the nutritional guidelines from the Eatwell Guide and the 'eight tips for eating well'. There are, however, a number of other food issues they need to consider, including a range of foods to avoid in pregnancy:
- Mould-ripened cheeses (e.g. Brie and Camembert), unpasteurised cheese (e.g. Stilton), pâté, pre-packed salad and 'cook chill' meals as these foods may contain *Listeria* bacteria, which can increase the risk of miscarriage or stillbirth.
- Raw and lightly cooked eggs, and foods containing them (e.g. mayonnaise), as these foods can contain *Salmonella* and cause food poisoning.
- Raw and partly cooked meat, unpasteurised milk and unwashed fruit and vegetables could infect the body with toxoplasmosis, which can cause flu-like symptoms in the mother and damage the nervous system and eyes of the baby.
- Foods containing high levels of vitamin A, for example liver and liver products such as pâté, which are toxic in excess and can cause birth defects.
- Some types of fish (e.g. swordfish and tuna) should be restricted in pregnancy as they may contain mercury, which can affect the nervous system of the foetus.
- Caffeine in tea, coffee and soft drinks can affect the absorption of nutrients. Pregnant women are advised to consume no more than 300 mg of caffeine per day.
- Alcohol can cross the placenta to the foetus, affecting its development. The Department of Health states that it is not possible to identify a safe level of alcohol consumption during pregnancy.

10 Priority health issues

1 Green leafy vegetables, pulses, fortified cereals.

2 In type 1 diabetes, the body cannot make any insulin. Type 1 diabetes occurs more commonly in children, adolescents and young adults. It is also known as insulin-dependent diabetes. The cause of type 1 diabetes is uncertain. It is an auto-immune

condition. In type 2 diabetes, not enough insulin is produced, or the insulin produced in the body does not work effectively. Type 2 diabetes tends to develop gradually as people get older – usually after the age of 40. It is also known as non-insulin dependent diabetes. The risk factors for type 2 diabetes are much clearer.

3 Answer to include **two** from:
- increased consumption of energy-dense food high in calories, fat and sugar
- decreased levels of physical activity
- overconsumption of fast food, processed food and treat foods
- larger portion sizes
- increased snacking and grazing
- high intake of sugary drinks
- comfort eating
- lack of physical activity
- sleep deprivation
- genetic traits such as a large appetite
- medical conditions linked to hormones.

4 Answer to include **three** from:
- eating nutritious foods and avoiding snacking between meals
- consuming foods such as milk and dairy products, which may protect against dental caries since they appear to reduce acid in the mouth
- drinking water to increase the flow of saliva, which neutralises the acid produced by plaque
- eating foods rich in fibre as this stimulates the production of saliva, which neutralises acid in the mouth and encourages remineralisation
- using fluoride toothpaste to protect teeth against decay
- visiting the dentist regularly – preferably every six months
- ensuring good oral hygiene by making sure that plaque is removed from teeth by brushing correctly and using dental floss.

5 Answer could include:
- saturated fats should be replaced with monounsaturated and polyunsaturated fats
- the essential fatty acids omega 3 and omega 6 are particularly important to protect against cardiovascular disease
- adults should not exceed 6 g of salt per day to minimise the risk of cardiovascular disease
- soluble fibre from grains, pulses and fruit can help lower blood cholesterol levels
- fruit and vegetables are a valuable source of the antioxidant nutrients vitamin A (beta carotene) and vitamin C, while wholegrain cereals are a good source of vitamin E
- maintain a healthy weight
- limit intake of alcohol
- be physically active
- avoid smoking.

11 Being an effective consumer when shopping for food

1 Anyone who buys a product or uses a service, in the public or private sector, is a consumer. To be an effective consumer means knowing about your consumer rights and responsibilities, understanding how to find expert consumer advice, being able to deal with issues such as poor service or faulty goods confidently and generally making your voice heard.

2 Answer to include **two** from:
- access – some consumers cannot access the goods and services they want or need because of a physical barrier associated with their mobility, sight, hearing or learning difficulty/disability
- age – this can be linked to judgements about a consumer's knowledge, confidence and experience
- ethnicity – language barriers and cultural differences can affect someone's ability to understand their rights and responsibilities as a consumer
- knowledge – a consumer's level of knowledge, numeracy and literacy will have an impact on their capacity to be an effective consumer
- resources – a consumer's effectiveness can be affected by where they live, as well as the time and money they have available and how they prioritise these.

3 Answer to include **three** from:
- the Northern Ireland Trading Standards Service promotes and maintains fair trading to protect consumers and enable businesses to succeed within Northern Ireland
- it enforces a wide range of consumer protection laws.
- it provides advice and guidance to both consumers and traders.

4 Answer to include **two** advantages from:
- may be cheaper than shops, particularly when buying in small amounts
- local and seasonal produce are promoted and widely available
- many markets offer an extensive range of food products.

One disadvantage from:
- packaging and labelling may not be available to help determine quality
- outdoor markets are dependent on favourable weather conditions
- markets usually only operate on specific days and early in the morning
- access can be an issue, e.g. parking in busy town centres where food markets are often located.

5 Online shopping is available 24 hours a day, 365 days a year. An extensive range of products or services can be purchased, including specialist and overseas brands. It is possible to price check a range of supermarkets, which can save money. Online shopping is convenient, with a choice of click and collect or home delivery service. A disadvantage of online shopping is that it is not possible to examine and select items in person, which may affect the quality of the food received. It is necessary to have a credit or debit card in order to pay for online shopping, which may raise issues of security. It can be time consuming to compare the prices and services offered by different retailers, and it is possible that items may be misrepresented. Often a delivery or return charge is made, which adds to the overall cost of the products purchased.

12 Factors affecting food choice

1 B: Informs consumers about the traceability of food produced in the Republic of Ireland.

2 Answer could include:
  - health concerns may cause people to exclude foods from their diet for medical reasons or to prevent disease
  - the influence of health issues on food shopping depends on consumers' interest in their own health and well-being
  - some consumers may choose foods that are labelled as being fortified with vitamins and minerals (micronutrients) because they wish to enjoy the health benefits that these foods offer.

3 Answer to include **two** from 'mandatory' and **two** from 'voluntary':

Mandatory	Voluntary
• Name of the food • List of ingredients • The quantity of certain ingredients • Net quantity • Indication of minimum durability (for example, 'use by' or 'best before' dates) • Storage conditions and/or conditions of use • Name or business name and address of the food business operator • Place of origin or provenance (if implied) • Food allergens • Nutrition information • Alcohol strength	• Nutrition and health claims • Front-of-pack labelling • Marketing terms, e.g. 'traditional', 'pure', 'homemade', 'natural' and 'fresh'. • Special dietary advice, e.g. 'suitable for vegetarians' or 'suitable for vegans' labelling

4 Answer to include **three** from:
  - supermarkets provide a clear space just inside the entrance to allow customers to adjust to the atmosphere; they may blow warm air on to customers as they enter the store, which makes them feel welcome
  - magazines or seasonal offers may be displayed at the front of the store, often to the right of the entrance, which encourages people to stay and browse before they begin the rest of their shopping
  - fruit and vegetables are often placed near the front of the supermarket as customers associate these with freshness and quality; placing at them at the front of the store has a positive effect on sales
  - bread and milk are essential purchases for most customers, so they are usually displayed at the back of the supermarket; this ensures that customers who wish to purchase them have to walk past displays of goods that they might be tempted to buy too
  - displays at the checkout are the supermarket's last chance to encourage customers to make further purchases; sometimes sweets are displayed, but many other types of goods will be displayed at the checkout, often depending on the weather or the season, e.g. wrapping paper is often located near the checkouts in the weeks before Christmas.

5 Food retailers offer a range of incentives that influence food choices made by consumers. These include price promotions such as a discount to the normal selling price of a product (e.g. 'buy one get one free' or '20% extra free') and money-off vouchers offering a current or future discount to consumers. Many food retailers now offer loyalty cards. These allow consumers to collect points every time they spend in the store, which can then be exchanged for money-off vouchers or other rewards, such as restaurant vouchers or theme park tickets. All major food retailers offer their own brands for a variety of food products and these are usually cheaper alternatives to branded products. Food retailers are keen to demonstrate that they offer value for money, so they often check their prices against their competitors and highlight products that they are offering at lower prices than other stores.

13 Food safety

1 Environmental health practitioner
2 Answer to include **two** from:
  - raw meat should be refrigerated and stored in the fridge at 1–4 °C
  - raw meat should be stored on the bottom shelf of a refrigerator

- it is important to follow extra instructions regarding how long raw meat can be safely kept in the fridge once opened
- bacteria will reproduce once food has defrosted, so thawed raw meat should not be refrozen
- raw meat stored in a freezer should be carefully wrapped and labelled.

3 Answer could include **two**:
- a 'use by' date mark appears on the labels of highly perishable or high-risk foods such as pre-packed sliced cooked meat
- highly perishable foods such as pre-packed sliced cooked meat go off quickly and so must be stored in a fridge or a freezer that is operating at the correct temperature
- pre-packed sliced cooked meat must be eaten by the use by date; after this date it is likely to become unsafe to eat and could cause food poisoning.

4 An environmental health practitioner is responsible for inspecting food premises to ensure they adhere to all food safety regulations. They enforce food legislation and establish food standards. The role includes advising and educating food-based businesses and delivering hygiene education. Another role is to investigate complaints about food, as well as to investigate outbreaks of food poisoning. An environmental health practitioner also responds to food alerts from the Food Standards Agency and checks that food labelling does not provide misleading information to the consumer.

5 When preparing, cooking and serving food, it is vital that you wash your hands regularly. This is especially important before preparing food and eating, and after handling raw foods, going to the toilet, touching waste, coughing and sneezing or handling pets. You should avoid handling food if you are unwell or caring for someone who is unwell. If you have any cuts, sores or burns, these must be covered with clean dressings or blue plasters, which should be changed regularly. Always wear clean clothes and a clean apron when preparing food. Remove jewellery before preparing food and avoid touching your hair. If you have long hair, it should be tied back.

14 Resource management

1 Answer to include **two** from:
- unnecessary and wasteful buying
- poor planning when shopping
- incorrect storage of food
- impulse buys.

2 Answer to include **two** from:
- choose own-brand food items as these usually offer better value for money

- prepare a shopping list to avoid impulse buys.
- price check between food stores to ensure value for money
- prepare and cook meals from scratch rather than buying convenience foods and ready meals, as these are often more expensive than homemade
- choose foods that offer good value for money
- compare unit prices to ensure you are getting the best deal
- avoid special offers unless they provide real value for money and the food can be used.

3 Answer to include **three** from:
- plan meals and write a shopping list to ensure you only buy what you need
- be accurate with portion sizes, e.g. a portion of pasta for an adult is approximately 100 g
- buy food with the longest use by date so there is time to use it
- make the most of food purchased by using leftovers for a second meal
- buy ingredients that are versatile and can be used in a range of different meals
- keep a range of useful store cupboard basics that can be used to create meals quickly, e.g. jars of herbs/spices, tinned tomatoes, pasta, rice, flour, couscous, noodles, stock cubes, beans, tuna, oils/vinegars
- store food correctly to keep food fresher for longer, e.g. place fruit and vegetables in the salad drawer of the fridge and store potatoes and onions in a dark cupboard
- freeze food before the use by date, defrost when needed and use within 24 hours
- always cook and eat first the food that was bought first. Newly purchased food should be stored at the back of the fridge, freezer or cupboard to make this easier to achieve
- avoid marketing strategies when shopping for food, which can often lead to impulse buys and generate food waste.

4 Using a contactless debit card to purchase a takeaway coffee is a quick and efficient method of payment. It can only be used for purchases up to £30 so is ideal for purchasing a takeaway coffee as it avoids the need to carry cash and to count out money at the till. However, there is some concern about the security of contactless payment methods, as it is not necessary to enter security information, such as a PIN. The customer's bank account must be in credit in order to pay for the takeaway coffee, as debit cards only allow consumers to spend the money available in their account.

5 Customers that buy food using a credit card may receive benefits linked to that card, e.g. cashback or rewards/loyalty points. Using a credit card to pay for food shopping avoids the need to carry a large sum of cash. Provided the customer pays

their bill in full at the end of the month, a credit card offers interest-free borrowing. Contactless credit cards offer convenience at the checkout as they speed up the payment process. However, credit cards are a form of borrowing and it can be a high-risk decision to use one to meet a basic need such as purchasing food. The customer should remember that interest will be applied at the end of the month if the bill is not paid in full and this increases the overall cost of their food bill and the risk of debt. There are also concerns about the security of contactless payment methods, because it is not necessary to enter security information such as a PIN.